M000275727

OLD BARNS IN THE NEW WORLD

Reconstructing History

Also from Berkshire House Publishers

Most Excellent Majesty: A History of Mount Greylock
Deborah E. Burns and Lauren R. Stevens

Shaker Woodenware: A Field Guide, Volumes I & II
June Sprigg and Jim Johnson

Shaker Baskets & Poplarware: A Field Guide
Gerrie Kennedy, Galen Beale, and Jim Johnson

By Ejner Handberg:
Shop Drawings of Shaker Furniture & Woodenware, Volumes I, II, & III
Shop Drawings of Shaker Iron & Tinware
Measured Drawings of Shaker Furniture & Woodenware
Measured Drawings of 18th-Century American Furniture

The Berkshire Reader: Writings from New England's Secluded Paradise
Edited by Richard Nunley / Drawings by Michael McCurdy

The Pioneer Valley Reader: Prose and Poetry from New England's Heartland
Edited by James C. O'Connell

OLD BARNS IN THE NEW WORLD

Reconstructing History

Richard W. Babcock and Lauren R. Stevens

Berkshire House Publishers
Lee, Massachusetts

728.9
B112o

Old Barns in the New World: Reconstructing History
Copyright © 1996 by Berkshire House Publishers
Photographs and drawings © by Richard Babcock or other
credited photographers and sources

All rights reserved. No portion of this book may be reproduced —
mechanically, electronically, or by any other means, including
photocopying — without written permission of the publisher. For
information, write to Berkshire House Publishers, 480 Pleasant St.
Suite 5, Lee, Massachusetts 01238 or call 413-243-0303.

Unless otherwise indicated, all photographs in this book were
taken by Richard W. Babcock. Every effort has been made to
determine the creator of the photographs included. Any claims to
the contrary should be addressed to the publisher.

Front cover, frontispiece: detail of a photograph by Richard Babcock
Back cover: photograph of Richard Babcock by Harding-Glidden

Edited by Deborah Burns
Cover design by Pamela Meier
Text design by Catharyn Tivy
Map by Karen Kane: page 27
Drawings by Jack A. Sobon: pages 52,54,57,62,88,97,99,101,102,
123,138,183

Library of Congress Cataloging-in-Publication Data
Babcock, Richard W.
        Old barns in the New World : reconstructing history /
    Richard W. Babcock and Lauren R. Stevens.
            p.   cm.
        Includes bibliographical references and index
        ISBN 0-936399-79-1
        1. Babcock, Richard W. 2. Carpenters — United States —
    Biography. 3. Barns — New England — Conservation and
    restoration. 4. Building, Wooden — New England — History.
    I. Stevens, Lauren R., 1938–   . II. Title
    TH140.B33A3 1996
    728'.922'0288
    [B] — dc20   96-21932
                                CIP

ISBN 0-936399-79-1
10 9 8 7 6 5 4 3 2 1
Printed in the United States of America

To Henry N. Flynt, Jr.

A Good Friend, a Trusted Adviser,
and a Generous Supporter

Table of Contents

A map showing places important to Richard Babcock
and his barns may be found on page 27.

OLD BARNS IN THE NEW WORLD

Reconstructing History

PREFACE

LET'S GO TO THE BARN.

Most children growing up today do not know what it is like to play in a barn. They don't know the musty, sweet smell of the hay, or the exciting, comforting sounds of large animals shifting their weight and rattling their stanchions. They haven't had the chance to marvel at cobwebs as thick as hawsers draped on ancient, mysterious, wooden wagons and leather harnesses in unused sections of the loft. They don't know what it feels like to swoop on a swing with impossibly long ropes from the dark inside of the barn to daylight outside, or the daring sensation of leaping down from the loft into loose hay.

They don't know the constant sounds and rustlings of barns, as the great wooden structures respond to the wind or the sun, as swallows crisscross under the roof beams, as small animals scatter at the approach of the barn cat. They are unaware of innumerable marks left by people and events of days and decades gone by.

Few of us alive today know the barn as the focus of long hours of hard work, whether emptying out wagons of loose hay and tossing it into the loft or milking in the winter, head against the steaming side of the cow. Many children and their parents do not know of the endless feeding of animals and forking out their manure.

Like great churches, the light slanting not through stained glass windows but through the irregular gaps between planks, barns make their occupants look up. As in a church, exactly what we see when we gaze into the vast upper spaces remains suggestive rather than definitive. Barns are homes of physical work in the Shaker sense: hands to work, hearts to God.

Not everyone enjoyed that life and fewer enjoyed it in its entirety. Much of it was drudgery. Yet barns spanned generations, in a time when children of farmers expected to be farmers themselves, continuing to work in the old way with the old tools. Today, small-scale family farming in the northeastern United States is in trouble. Because farms are no longer profitable, they aren't being maintained, and many are growing up into woodlots before our eyes, or becoming the sites of housing developments—or malls or highways.

It is not surprising that wooden barns are disintegrating as the minutes slip by or, more abruptly, are being bulldozed out of the way. Many still standing have been stripped of their siding to provide interior walls for apartments or mall stores. Yet barns depict a way of life and a period of history more accurately than any other structure, including homes. A drive on any road in a rural area illustrates the state of farming, by the fate of barns. It dramatizes the end of an era when generations in a family worked together, for better or for worse, earning their living from the land.

Maybe someone will be able to figure out how to save small-scale farming as an economically viable enterprise. Even that happy outcome would be unlikely to save wooden barns, however. The truth is that wooden barns outlived their usefulness long ago—before most of those we see with their roofs gone and their timbers sagging were even built.

Among wooden barns, precious few of which remain active parts of farm life, one type is especially rare and historically significant—those built by the first European immigrants to arrive, while the United States and Canada were still colonies. These Old World barns were built in the wilderness of the New World exactly as the settlers' ancestors had built them across the ocean. They were built so well that, surprisingly, some of them still exist 200 to 300 years later, mostly in New England, upstate New York, and Canada.

One man knows them better than anyone else, from the practical experience of taking them apart and putting them back together—and from studying everything he could find relating to barns. He has tried to save them and at the same time tried to make a living by converting them as homes. Whatever may be said about the congruity of those two goals, Richard Babcock's experience in moving 100 barns has not made him rich. Although he has attracted enormous attention in the press and television, and is acknowledged as a national expert, he still must struggle to save even the few oldest barns that are left.

Had Richard Babcock been twenty years younger or lived 100 miles away in any direction, he would know far less about Old World barns than he does. This distinctive and practical form of American architecture might have moldered as an historical or architectural footnote, without the general public having much more than a photographer's or painter's interest in it. This book seeks to take advantage of Richard's activities to provide a rounded view of these barns.

The time is past ripe to write the history of North American barns and their influence on the growth of this country. At present some history exists in books that primarily serve more decorative functions, some persists in the dwindling number of families who still live with roots barns, and some can be found in the accounts of early regional historians. Scant information exists in deeds and wills. With only one or two exceptions, the names of the architects or master builders of these noble edifices are lost.

The absence of written records and a tradition that emphasizes the house rather than the barn are the main reasons why we have little barn literature. Usually the barn is simply cast among "outbuildings." Yet the barn was a larger and more important structure than the house. It told more about the farmer and his worth. It usually predated any but the most rudimentary sort of human shelter. In fact, according to historian Charles Klamkin, the saltbox that housed so many of New England's colonial settlers derived from early barn design.

Not too long ago disaster struck the museum of ancient barns Richard Babcock was trying to create in western Massachusetts: a fire destroyed buildings and priceless artifacts. At the same time, he was reminded of his own physical vulnerability, ending up in the emergency room of the regional hospital.

Three of his friends met while he was still bedridden, to see what they could do to help. Richard had self-published two booklets on his work, which he hoped could get wider circulation. The friends made a suggestion, which Richard accepted, that he work with a writer to co-author a new book that would be about him and what he had learned about his beloved barns. Those involved in the project would like to believe that from the flames came some flickering good.

Acknowledgments

The Berkshire Athenaeum, *Albany Times Union, Berkshire Advocate, Berkshire Eagle, Berkshire Magazine,* Cobble Cafe, *Boston Globe, Country Journal, North Adams Transcript,* Gannett Westchester Newspapers, Horicon Newspapers, *House Beautiful,* Historic Deerfield, Inc., Historic Hudson/Philipsburg Manor, The National Geographic Society, *New England Monthly, New York Times,* The Sterling and Francine Clark Art Institute, *Time,* The Williams College Libraries, The Chapin Rare Books Library, Williamstown Public Library, *Washington Post,* Wolf Trap Foundation for the Performing Arts, *Yankee*... Alice Babcock, Clayton Babcock, David Babcock, Donna Babcock, Margaret Babcock, Deborah Burns, James M. Burns, Mary Jo Carpenter, Lewis Cuyler, Peter Erickson, John and Josie Eusden, Henry N. Flynt, Jr., Wayne Hammond, Edith F. Howard, Karen Kane, Benjamin W. Labaree, Jr., Judith Leab, Philip Rich, Jean Rousseau, Catherine Filene Shouse, Michael Steele, Theodore Sylvester, Marc and Micheline Toureille, Dustin Wees, Lawrence Wright... and many others, over many years.

Richard would particularly like to express his love and appreciation to his sons and daughters, who worked with him, and to old Henry, Jack Sobon, Paul Martin, Richard Deslauriers, and William Flynt.

I

FIRE!

ON THE EVENING OF JULY 5, 1994, fifty-nine-year-old Richard W. Babcock of Hancock, Massachusetts, was watching television and relaxing after a long day of dismantling and reassembling a post-and-beam barn. His wife, Donna, was sleeping, tired from taking business courses. When he got up for a drink of water at 11 p.m., he spotted a glow out the kitchen window.

"My God, the barn is on fire," he said out loud and then yelled upstairs to Donna, "Call the fire department!" Donna awoke quickly to make the call.

In previous years this bear of a man had moved and reconstructed three barns, of a proposed seven, in a grassy area behind the house to form the beginning of a museum of barns. He had erected a German carriage barn, circa 1750; a smaller Dutch barn, dated to the same year, that he used as a workshop; and his pride, the Great Plantation Barn of Saratoga, 1685.

← **Richard has sliced a cross section from one of the charred beams of the French barn, stored in the Saratoga barn that burned in July 1994, to see how much of the wood is still good.**

In the glow Richard could see his wire-haired terrier and his shepherd/hound cross, one penned next to the ancient Dutch barn and the other tied a few feet

15

The timbers of the Great Plantation Barn of Saratoga after the fire of 1994.

away. He tore open the door and heard them barking. Richard ran like the dickens the 150 feet to the barn, grabbed one dog, pulled it over the top of the pen, and then fumbled to unsnap the other.

The fire had fully engaged the Saratoga barn. Inside it were the timbers of a French barn he had found moldering on a farm near the Vermont border in New York State, a barn he ascribed to 1540—a date that would push back settlement in the interior New England and upstate New York region to eighty years before the Pilgrims landed at Plymouth Rock. He had hoped to re-erect it for the museum when he had funds and time.

The Saratoga barn also stored timbers from a small barn he had framed for display purposes, something he could, with help, raise in one day—as he had done on Williamstown's Field Park a year before. The barn contained the only known extant hewn poles of a hay barracks, a way of storing hay in the

16

field before barns were available, and a mammoth apple grinder that was built to be turned by slaves. These were loved structures, his livelihood, and potentially important historical artifacts.

Hearing sirens, he ran to the Old Hancock Road, now replaced for most traffic by Route 43, to meet the first volunteer firefighter to drive up in his pickup, Fred A. Scace, Sr. Then Richard felt pains shooting through his body. He lost his breath. He had had such pains—angina—once before. He knew that he shouldn't let himself get excited.

But how could he help it? All he could hear was the sound of timber crackling. Fire Chief David Rash says the flames leaped more than 30 feet high from the 50-foot by 50-foot structure. Richard knew that first one fireman and then several were trying to calm him down. "I felt like Samson after his haircut," he remembers. "I felt I had lost my strength. The most important stuff I had was in that barn."

The firemen called an ambulance, which took Richard to the Berkshire Medical Center in Pittsfield. He nearly died on the way and spent a week in the coronary care unit. The barn, and what was inside, burned to a few charred timbers.

The Setting

Hancock is an out-of-the-way town in an out-of-the-way northwestern corner of Massachusetts. It is shaped like a long, north-south rectangle, and its mountainous terrain makes it impossible to drive from the north end to the south end without leaving town. Richard lives in the north end, considered the town's center or "village." In the south end, Route 20 runs east to Pittsfield and west toward Albany, New York, passing through Hancock Shaker Village. In between the north and the south end lies Pittsfield State Forest, crisscrossed by trails but with no roads passable by automobile.

Hancock was called Jericho Plantation in its pre-Revolutionary days, before being renamed for the signer of the Declaration of Independence. Jericho it remains, walled by steep mountain sides now containing two downhill ski areas: Jiminy Peak Mountain Resort, in Hancock itself, and Brodie Mountain, on the east side of a ridge that helps form the valley. To the west the Taconic Range borders New York State; Williamstown is neighbor to the north; New Ashford, Lanesborough, and Pittsfield lie to the east, and Richmond, to the south.

Berkshire County, home to Hancock, Williamstown, Pittsfield, and other Massachusetts communities Richard frequents, is perhaps the only county in the United States that borders three states—New York, Vermont, Connecticut and, some say, a fourth: Massachusetts, implying that it is somewhat forgotten by those who live and govern 150 miles to the east, in Boston. An area where dairy farms still struggle for existence, the county has seen most of its major industries reduced, so that it depends increasingly on tourism. It is different from the suburbs around Boston, which once were farmland. Even though Massachusetts is a small state, this county, nestled among the Taconics, the Green Mountains, the Hoosacs, and the southern Berkshire plateau, feels remote.

Historically its north-south ridges and rivers made Berkshire County easier to get to from Connecticut than from Boston, which partially explains why most of its earliest residents came from the south and why Williams College, located in Williamstown, has always had closer ties to Yale than to Harvard. In fact, the county claims a dual allegiance to Boston and New York City. Some root for the Red Sox and some for the Yankees or Mets. Most of the seasonal visitors are from the New York metropolitan area.

Berkshire County is hilly, with deep valleys, most of the ridges still tree-covered rather than dotted with second homes. The larger towns cluster in the valleys; much of the open land in the hills is publicly owned state forests and parks. Easily within Richard's circuit to the north are the dairy farm hills of Vermont, to the south the old settlements of western Connecticut, and to the west, the elaborate delta formed as the Mohawk River from western New York State joins the north-south Hudson River.

At that junction rises the capital of New York, Albany, where 300 years ago the Dutch and then the English traded tools, cloth, and guns to the American Indians for furs. The Hudson Valley and, to a lesser degree, the Mohawk were early owned by wealthy patent-holders who enticed settlers to the rich soil of their shores.

"Barn Poor"

Richard's life work was the discovery of the "roots barns" of America. By taking apart and reconstructing more than 100 barns, and researching the ownership of most of the barns he worked on, he taught himself which designs and builder's techniques stemmed from which nationality. It is Richard's genius to be able to tell by the construction techniques the date and nationality of the builders of the earliest barns in this country. He came to know which were built by the Dutch, which by the Germans, which by the English, and which by the Scots-Irish, who settled in New England and eastern New York prior to the American Revolution. And, bucking historical tradition, he believes he had evidence that the French settled here, long before the others.

Many of the structures were raised with the help of slaves. Richard's research repeatedly convinced him of the importance in New York and New England of African slaves, who performed the backbreaking work of clearing the fields, building the massive barns, and maintaining the farms.

Richard Babcock had been developing a museum to show the variety of barns first-generation Europeans put up when they settled here. He had also written up his experiences and knowledge of barns, self-publishing two books, the second a variation on the first, both full of the photographs he took to document his efforts. The heart of his museum, and some of the most important evidence for his historical theories, turned to ashes that night.

By the time of the fire, Richard was nationally recognized as an authority on early barns. Furthermore, a resurgence of interest in timber-frame construction owed much to his skills and the people he had trained. He had made his reputation, but not his fortune.

He may have made some enemies, as well. State Trooper Christopher Ware later said, as a result of his investigation, that "human hands were involved." There was no electricity and no baled or wet hay that could have spontaneously combusted. Ware found some evidence: fireworks scattered outside the barn. The investigation did not turn up any culprits, however, and the fire marshal determined that no accelerants, like gasoline, were in evidence.

Richard cannot believe that the ancient barn was deliberately set on fire,

Richard had restored the Saratoga barn
to museum quality, including a hay rack
and an apple grinder originally built to
be turned by slaves.

saying, "I just hate to think someone would do that." Besides, the dogs would have alerted him if anyone had been in the area that night. About a week before, in fact, in the evening, the dogs had raised an alarm. Richard went outside and heard a splash in the creek that flows through his property. He found no one, nor did he ever figure out what made the splash. An animal, perhaps.

Now he thinks it possible that some small spark from kids playing with fireworks the previous day, July 4, caught on some of the timbers and smoldered around the clock. Perhaps the dogs barked then at the kids, but neither he nor Donna were around at the time. Or perhaps, as on the evening of July 5, the television was turned up too loud to hear the dogs.

Lying in Room 427 of Berkshire Medical Center, Richard said he was "trying to figure where's the good that comes with the bad." He had a history of capitalizing on misfortune, like the time he was laid low with lead poisoning. Because of the enforced time of taking it easy, he had turned to early records of land transfers, a study that helped him formulate his historical views of barns.

At BMC now he added, "I loved my barns. I can't imagine I'll never see them again." Most of his life had been devoted to barns. "I don't want to quit," he said. People came to see him and called him on the telephone to wish him well. He appreciated that.

As the numbness of the tragedy wore off, he was left in emotional pain, especially as he thought about the lives that were connected with his barns. "The dream is still strong," he said from his bed, "but I am not young any more."

He was on medication. After he left the hospital, he felt drugged for weeks, so he altered his medication to rely on dieting and aspirin. Richard came to two conclusions from his heart attack: He wanted to continue training young people as he had done with apprentices who had come to work for him, and he wanted to find an editor and publisher to give wider circulation to his self-published work.

After leaving BMC, he amazed his friends by returning to work immediately, although he "couldn't do a hell of a lot." He did not have the financial resources for a leisurely recuperation, let alone retirement. "Barn poor," he always calls himself.

Although he had received reams and reels of publicity over the years—

three stories in *Yankee Magazine,* stories in *New England Monthly, Americana, National Geographic* News Service, and *Time Magazine,* and television coverage of barn raisings at sites like Wolf Trap; although he was recognized in the U.S. Postal Service write-up of a commemorative stamp; and although he erected other public barns like those at Philipsburg Manor, Macomber Farm, and the Animal Rescue League of Boston's Dedham farm—still he had accumulated no nest eggs in his barns. In fact, his major tangible assets had burned. Richard puts more into his barns than he takes out.

No Stranger to Controversy

Steeped in the skills of post-and-beam construction learned the sweated way, Richard is at once the most knowlegeable person anywhere about barns built in this country before the Revolutionary War and, by great good fortune, a resident of one of the few areas of the country where some of these endangered species still stand—reduced in number each year.

Over the years, from what he learned from his grandfather and from his own hands-on studies, Richard has developed his understanding of barns, how they were put up and how they were taken down. Some of these beliefs are controversial, within the rather small community of those who are interested in barns, post-and-beam construction, or the skills of joinery. Even Richard's career of saving barns by taking them down, moving them, and reconstructing them elsewhere, usually as homes, has its detractors. Richard responds to that criticism by agreeing that in a perfect world all the Old World barns would be saved and continue to play an active role in a vital farming community. In an imperfect world, it is better to move barns that would otherwise rot. They will then continue to be appreciated, whether as homes or as barns again.

Richard's education in barns began when he was young—and when the most precious barns were already ancient.

23

I I

"ROOTS BARNS"

THE WORD *BARN* COMES FROM two Middle English words meaning "house
for barley." In this country, the kind of barn we are interested in served best as
a "house for wheat." Since they were built, their practical uses have diminished.
Having grown fragile themselves, they continue to tell the story of how their
builders grew a country against great odds.

The history of barns is the history of the earliest, bravest, and most fragile time for Europeans in this country. Those settlers who survived crossing the
ocean stepped off their reeking ships onto land with great promise but only the
barest of American Indian clearings. They built barns, as soon as they could,
to hold their precious food, even to live in for a while, and to obtain some degree of permanency on a land they saw as hostile. But soon they built barn edifices to which they could cling when all else seemed in danger of washing away.
Plymouth Rock was a stepping stone compared to the importance of the rocklike
barns on which settlers built their livelihoods. The colonies grew like small
shoots, ever stronger, over these barns.

They were built to last for centuries, but changes in agriculture made them
obsolete within generations. They were built to hold wheat, but farmers' greater

need was to shelter animals and their feed. In the twentieth century, changes in the dairy industry made the ancient barns a burden to farmers. For some time they tried to adapt post-and-beam barns to new uses; in the last fifty years that has become impossible. Of those early barns, built by the first generation of Europeans to settle here, no more than a handful still exist as elements of working farms.

No directory of these barns exists, except perhaps in one man's head. Richard Babcock's list of Old World barns built in the New World includes a remarkable variety of structures built in Massachusetts, Connecticut, upstate New York, and New Jersey. He calls them "roots barns," because they tell us of our heritage from the Old World, and of our brave beginnings and fragile shoots here. They include the original barns built when the seacoast of New England was first settled, as well as barns built inland.

Regional Barns

Richard's inventory does not include whatever barns were built in the South prior to the Revolutionary War, most remains of which were destroyed during Sherman's March in the Civil War. Nor does it include barns built in Canada, although Richard has visited many of them and for the most part they are the same types he has seen in his backyard. Canada seems to have a larger number of extant samples than does the New York–New England area.

The barns that grew in the Midwest and West did not belong to the earliest period, although those farmers began by replicating the eastern structures with which they were familiar. Changes in building materials and techniques, the drastically increased size of western farms, and the unprotected exposure to the wind altered barn design in that part of the country.

His inventory and this book do not focus directly on the distinctive style that arose in Pennsylvania, influenced by Swiss Mennonites who moved first to Germany and then to the New World, only to be misnamed Pennsylvania Dutch. Their barns, which were part ancestor to second-generation barns in the New York–New England area, were built on a slope, as were all barns in Switzerland, with the livestock living in the basement.

25

So practical was this approach that where topography didn't present an incline, industrious Mennonite farmers built a small hill and constructed a ramp up to a second-level door. In this way they could wheel their hay directly from the field into the second level of the barn, where it was stored, and easily dispense it to the animals at ground level.

More northerly farmers at first kept hay and animals on the same level or figured out ways to hoist hay upstairs to the loft. Increasing settlement, however, forced them to move away from the bottom lands along the rivers, with a greater emphasis on animals. In the uplands they were often presented with a suitable hill into which to set the barn, so they too built a lower story—or at least a ramp.

Almost all of the architectural features of the European barns that took root in colonial America came directly into the Northeast, the territory with which Richard is most familiar. Within his lifetime the last surviving examples of those barns have become redundant, a burden on the farmer or an impediment to the developer. This is due to the conversion of most family farming in this region into dairying, along with changes in the dairying industry itself. The first surprise is that any Old World barns still exist at all; the second, more sobering shock is that a decade or two may finish off the last of them.

Yet the roots barns stand—those that still do—as extraordinary monuments. Built of the timber available from clearing the fields, the early barns of this country lacked excessive detail or any parts that were not completely functional. All were built of timber frame—post and beam—covered with roof and walls.

Unlike homes, factories, or even barns' close relatives, churches, in which the frame is at least partly hidden, the way barns are put together is visible from the inside. A barn is, therefore, an honest and straightforward solution to storing and sheltering, as has been noted by John Fitchen, an authority on New York State Dutch barns.

Ancestors of Barns

The roots of barns go back to ancient agricultural civilizations, such as that of the Egyptians. Joseph, after all, won fame for helping Pharaoh store in something like a barn the excess wheat of the seven fat years against the seven lean

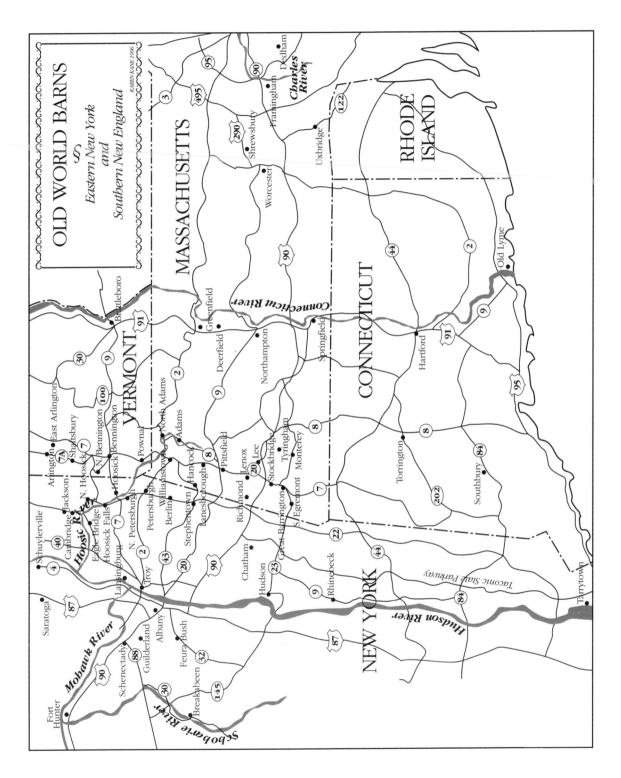

years. Cycles of plenty and want were customary in the Middle East and must have made suitable storage structures necessary.

Barns the shape and structure of overturned ships were built in Iron Age Britain, as Eric Arthur and Dudley Witney have noted in their book *A Vanishing Landmark in North America*. Furthermore, the "cruck framing" then used — using curved trees to form pairs of posts from the ground to the ridge — was similar to boat construction, and the ancestor of post and beam.

Among the Saxons and other early farmers of Europe it became a custom to house people, hay, and animals under one roof. Examples of such structures still exist, although they have been converted to non-barn uses, in Europe and Canada. The male servants slept over the horses and the women slept over the cows, according to Arthur and Witney. We know that at least one combination barn, animal shelter, and home was built in what was to become the United States. Hudson River patent-holder Killian van Rensselaer ordered one for one of his tenants.

Both the functionalism and the disadvantages of such a system are easy to imagine. One structure is easier to build than two or more. The farmer could tend to his chores in nasty weather without experiencing it; his wife could watch over the valuable livestock without leaving her weaving or cooking. In theory, the fire that the family cooked over and that heated them helped to warm the animals — although given the inefficiencies before fireplace or chimney, the body heat of the animals, trapped by the hay insulation above, probably helped to warm the family more than the fire warmed the animals.

Fire is one of the great dangers of this system, especially as the roof was likely to be thatched. One errant spark could quickly destroy everything the farmer had, whereas the loss of a separate house might leave untouched his most valuable goods. The same issue arose in nineteenth-century Maine, when home and barn were connected. Because of fires used for light, heat, and cooking, homes were more susceptible to burning down than barns, but took the barns with them if they were attached. Several Down East towns passed ordinances prohibiting barns and houses from being joined.

Early Europeans were far less fastidious about odor and hygiene than we

28

are; nevertheless the aroma, insects, and vermin attracted to a combination barn/house cannot have been pleasant. Perhaps that was another reason the common dwelling did not catch on in colonial America.

The rise of the church in Europe in the Middle Ages inspired some of the cathedral-like barns that still exist—in fact, barn and church architecture remained similar in the New World. The oldest timber-frame barn still standing in Britain is the grange barn in Coggeshall, Essex, estimated to have been built around 1140. It has been restored by the National Trust. One of the grandest barns is Great Coxwell, in Berkshire, England.

The abbeys provided the labor pool for the great church farms, which demanded enormous storage. In addition, the church taxed large landholders to give a tenth of their plenty to the church. Thus the tithe barns were built to hold that plenty.

By the time of the Spanish Armada, 1588, Europe was running out of the kind of massive trees necessary to build huge wooden barns. In fact, for some time barns with partial stone walls or mud and straw walls had been becoming more common. (In colonial Pennsylvania, where there was ample wood, the Swiss Mennonites continued to build with stone, mud, and straw, illustrating an important truth of barn construction: it is a highly traditional art, slow to adapt to new conditions.)

As a result of the attack by the Spanish Armada, England forbade the cutting of the last of her largest trees, reserving any usable timber over 16 feet for the Royal Navy. Before carpenters journeyed to the New World, the size of timbers they could build with in the Old had been reduced. Thus timbers in seventeenth-century European barns seem relatively small and curved. Despite the labor of hewing timbers for barns, the master builders from the Old World must have rejoiced at the enormous, straight logs they had to work with when they arrived in the New.

Master Builders in the New World

In North America they found endless forests with towering trees, some rising directly from coasts and rivers. They found and appropriated clearings left by the American Indians, starting points for settlements the Europeans soon out-grew. They girdled or burned or hacked down trees in order to plant crops. That wood—and much more—was available for construction. And upon the sides of those hills grew tall, straight trees, the likes of which they had never seen.

In most of the colonizing efforts, master builders were deliberately sought out to accompany the settlers. Contrary to some sources, the farmer did not de-sign and supervise construction of his own barn here. We know this in part be-cause of the remarkable similarity of building techniques, based on the mortise and tenon joint, even across nationalities. The primary reason that we know master builders were in charge, however, is the thorough professionalism of the job. Roots barns were not built by gifted farmers who temporarily turned their hands to engineering. As Richard has recognized, they were built by carefully trained experts, with access to skills that were deliberately kept secret from those not members of the guild.

The master builder selected the timber. In some cases, the farmer cut it and seasoned it. Other workers may have come to do the hewing. Under the master builder's direction, the farmer, often with the help of his slaves, prepared the site. Generally the master builder laid up the dry stone foundation and the sills. The master builder laid out the bents, or units of the frame. Each may have been fitted on the ground in at least two separate ways, as Fitchen surmises, to test how each timber fit with its fellows.

Raising a Barn

Generally the master builder and his assistants put the barn up one post at a time. Sometimes, when the bents were fastened together, the farmer invited all the able-bodied men he could locate for a daylong barn raising, to do the heavy lifting of getting the elements into the air. The women came, too, to prepare the massive quantities of food and beverages all would consume. The master builder orchestrated the event.

If the raising was on schedule, by the end of the day the frame was in place, including the rafters. Occasionally the raising included part of a second day, but all parties knew the volunteers were anxious to return to the work of their own farms. Thus the master builder was judged not only on the barn itself but by the speed and efficiency with which the parts he had laid out went together.

The master builder and his help finished the roof, the boarding, and the doors. Sometimes the farmer may have taken over in the latter stages, but Richard believes that would have been a rare case, because the master builder wanted the entire barn to bear his signature.

That was the most colorful and publicized approach. As recreated, it is wonderfully photogenic. As Richard has demonstrated, however, using the right kind of ancient hoisting equipment it is possible for two or

A detail of the joinery in a roots Dutch barn that rotted into the ground in Altamont, New York.

three men to put up an entire barn. Generally that was how it was done. It may be that some communal groups, such as the Pilgrims or the Mennonites, more frequently joined in to help one another than average farmers did.

To the degree that barns were overbuilt, with oversized beams and an abundance of braces, the builder's motive was prudence, not showing off. Barns were built that way because no one could be absolutely sure what size timbers would be sufficient. Compared to their contemporary European counterparts, American barns were very large, because the timbers here were long enough to span large spaces. Because the spaces were huge, it was better to overbuild. Furthermore, given the size of the trees available, it was easier to use large beams than to hew the large ones down.

Contracting with Master Builders

We may wonder how farmers could rely on the timely arrival of master builders, considering that often their farms were widely spread out in areas that were nearly wilderness. The early Dutch farmers, along with some Germans, came to New York State at the behest of manorial lords who owned vast tracts of the Hudson and Mohawk valleys. The holders of the patents provided the master builders for tenants located on the best soil, where the crops were unlikely to fail. These skilled craftsmen, their wages paid by the lords and charged against the farmers' accounts, traveled from farm to farm as the need arose. For the patent-holders, providing a good barn was a form of insurance for the safety of the crop that would enlarge their wealth.

In areas of New England and New York where there were no such manorial lords, the farmer more often went without a real barn in the first few years, making do with hovels and sheds for his animals and storing his hay outside. We know that when the earliest Europeans arrived, many from the milder, southern end of England, they did not anticipate the need to protect their animals from the the long, bitter winters. Many beasts died.

Even where no lord of the manor made the arrangements, farmers waited for itinerant journeymen, who went where they heard they were needed, all the time training a few especially worthy assistants. No doubt farmers often had to put up with great inconvenience in the meanwhile, especially in storing wheat.

How did these architects/engineers/builders get paid? We simply don't know. We do know that the men in charge of these architectural masterpieces that still stand after 250 to 300 years, to which we give the humble name barn, took enormous pride in their work. It is written in every joint, every mark of the broadax or adze—even if the master builder's name is lost.

Of course it is possible that the barns that have survived and been recorded were the ones for which construction was professionally overseen—and that the barns that expired long before the twentieth century were built by the farmer. Of barns that we know about, however, none seem to have collapsed simply because of age. The greatest threat was—and, as Richard's experience testifies, still is—fire. The second greatest threat was lack of maintenance. If the roofs weren't kept up, even if the doors were left open, sooner or later the great timbers began to rot. Like an old dog, in motion so slow that it lasted over years, the barn settled to the ground, one end at a time.

Flatland Barns

Most of the first-generation barns built anywhere in the region were what Richard calls

This photograph of an early German barn that Richard moved and rebuilt shows the mortises that received the poles on which the wheat was stacked to dry.

flatland barns. They were built along the rivers, often in fields that had already
been cleared by American Indians. There the farmer grew wheat to make bread.

Wheat was cut early in the field, before it became so dry it would fall off
the stalk as it was being cut. After being gathered into small bundles, it was
loaded by hand into a wheat cart and brought to the barn. There it was stored
on top of a layer of loose hay spread over poles that lay over the beams. Richard has found some of these poles in a few ancient barns. The wheat was carefully stacked like firewood, the heads never touching the other piles. Tons of
wheat mounted to the ridge.

When the wheat was dried, it was dropped to the threshing floor below.

The planks there had spines between them to prevent wheat sifting through. The main floor of the barn was used for threshing, the large, opposing doors allowing the prevailing wind to separate the wheat from the chaff. Thresholds held in the wheat. The farmer threshed the wheat with flails, two sticks fastened together with rawhide. He picked up the straw with a fork to use for bedding. The grain was stored in barrels to be ground, early on by hand and in later years at the mill.

During the threshing the farmer had to pile the unclean wheat in mounds on the threshing floor, waiting for the right wind. The prevailing wind, toward which the barn doors were oriented, usually arose in mid-afternoon.

Farmer David Stiles of Southbury, Connecticut, told Richard of Stiles's father's desire to teach his young son how to thresh by hand. Stiles's family barn was built in 1673. Even though no good reason existed, his father persuaded Stiles to plow and plant a field with wheat. After harvest they bundled the wheat and stored it high in the barn. Stiles still has bundles of wheat from that time on a rack overhead in his barn. As Richard tells it in his own words:

Dave said he could remember well the day his dad said, "It's time to do the threshing, son." Dave brought the bundles down, spreading them on the floor. His father got out the flail, saying, "Watch me closely—pay attention." His dad brought the sticks back over his shoulder, the second trailing the first, then brought them back over quickly, striking the wheat.

Dave tried it. At first it went well, until he got tired and started to slow down. "That's what my father was waiting for," explained Dave. On a slower swing, the swingle or second stick hit him in the back of the head. "When I yelled, Dad said only, 'There, I told you to bring the stick over quickly.'

"Dad just wanted to see me hit myself on the head. It worked. I learned."

The rest of the time the big, main floor was used to store farm equipment, to tend to tasks in poor weather, to hold church meetings or even dances. It is said that after years of threshing, the floor became as smoothly polished as the dance floor in the local tavern.

The American Barn

Farmers were eventually crowded out of the rich bottom lands necessary to sustain wheat crops and moved into the higher land on the valley walls. There they were forced to change the focus of their farms to animals, which they could raise where the soil was not so good.

Due to the ingenuity of the post-and-beam construction, barns could be disassembled and moved with the farmers. A master builder wasn't always responsible for the disassembling and reassembling of the timbers, necessary unless the move was short and the barn could be towed by oxen. Richard has found some barns whose replacement parts installed at the time of the move were inferior, suggesting that the farmer called in barn movers rather than a master builder. He has spotted some barns that were moved three times.

In fact, even if a farmer had stayed in the lowlands, he would eventually have exhausted his soil for wheat. The only way farmers could maintain wheat production before chemical fertilizers was to keep moving to new, naturally fertile land farther upstream. Most farmers in this area changed from wheat growing to dairying or sheep raising soon after the Revolutionary War. The minority who were still in wheat in the nineteenth century lost out when the railroads began chugging into town loaded with wheat grown in the rich topsoil of the West. Western wheat was cheaper than the local wheat even with transportation costs added in.

The farmer had always had animals in the barn—a pair or a team of horses, a cow, oxen, chickens—but increasingly he turned to the animals to provide salable milk, butter, and cheese, sustained by the hay and later the corn he grew for them. Plenty of people, especially those in the enlarging cities, were ready to buy dairy products. So it was convenient to borrow from Pennsylvania barns by moving the old barn onto a basement where the animals were sheltered. Often the original orientation was changed, especially to get the side of the basement open to the air facing south for the warmth of the winter sun.

The barn now became a place not primarily to store wheat but to shelter livestock and their food, and feeding them was facilitated by gravity. A ramp or the side of the hill led to the main floor, so that a hay wagon could enter and its load be stored. Thus the original roots barns and the newer ones gradually were modified; the distinctive national traits merged. Barns became American.

Farms Evolve

The idea that all early farmers grew everything they needed to live and nothing to sell, is an exaggeration, as Howard S. Russell in his *A Long, Deep Furrow* has shown. Although homeowners, regardless of occupation, kept a vegetable garden, they also depended on commercial growers. From early on, ships and then roads were available to move food to market. Excess beyond what the farmer needed was taken to more populous centers. The dairy farmer converted that portion of the milk he was selling into butter and cheese until refrigeration made it possible to store fresh milk, at least for a few days. The Maine Central was the first railroad to use refrigerator cars, in 1881, allowing farmers to send milk to the cities a hundred miles away. The barn was at the core of commercial farms.

No evidence exists that a significant number of barn/houses were built in colonial days, as was the ancient European tradition. Instead, the earliest farmers tended to build several outbuildings while still living in a substandard house. The rule of thumb was to bring in at least seven or perhaps nine harvests before turning to build a substantial home. As the barns developed, other farm functions were combined under one roof—animal storage, wagon shed, milk house, and so forth.

Eventually the harsh climate of northern New England, where in a blinding snowstorm a farmer doing his chores could get lost a few feet from his back door, led to the "big house, little house, back house, barn"—the connected layout of a familiar nursery rhyme. The big house was the family home, the little house was a kitchen wing, the back house was either for wagons or a way of referring to the privy. Animals and their feed resided in the barn.

With the phasing out of slavery in the North, the opening of the Erie Canal in 1825, and the gradual exhaustion of soil in the Northeast, many farmers began to head west to try to do better what they had been doing here. The move was abetted by tariffs and by the freak cold summer of 1816. Farmers, especially those on marginal soil in the hills, were struggling—in case we think that is a late twentieth-century phenomenon. The best farms, generally those in the valleys, continued on and even enlarged, incorporating the land of those who left. Generally it was the economically marginal ones that went out of existence.

Farmers were profoundly conservative, not revolutionary. They went west not looking for change but hoping to be more successful in what they did. Nevertheless, the westward movement of cattle, sheep, and then wheat helped prepare for an agricultural revolution. Change came, even if unsought.

At the time of the Civil War, farming became commercial, not because it seemed desirable, but because it was necessary. Farms faced a shortage of hired help because of the end of slavery, the migration west, and soldiering. Consequently they turned more completely to machinery, which required capital. Thus farmers became wedded to their banks and to their cash crop. Many, of course, did not really know what was happening, leading to personal tragedies that continue to be played out today.

Hay by itself was not enough. By the middle of the nineteenth century dairymen discovered that if they fed their cows corn and grain they could milk them year-round. Furthermore, they could avoid the "spring flush," when green grass began to grow and the cows gave so much milk it flooded the market. Farmers stored corn in cribs (and still do). Then they began to chop green corn with stalks or green clover into ensilage for winter feed. Starting in the 1870s, according to Russell, this fodder was stored in silos, designed to keep out the air to prevent fermentation or rot.

Writing in a guide to farm buildings first published in 1881, Byron D. Halsted notes that the word silo comes from the French for "pit." Farmers in Europe had been experimenting with different ways of preserving green fodder into the winter by burying it. The new idea was to borrow the principle of preservation used in canning fruit—keeping out the air—by means of tight build-ing construction and by packing down the fodder. This change meant that the traditional barn was no longer as necessary for food storage, whether for ani-mals or people, as it had been.

Pole Barns

The traditional post-and-beam construction of barns, houses, and ships was also about to undergo a major change. Lumber mills started turning out two-by-four studs and lumber of other dimensions, with straight edges, and quickly revolu-tionized house construction. The change was slower with barns, which contin-ued to be hung on massive timbers for generations, even if the siding was nailed to sawn studs. For one thing, "new" barns were often recycled old ones, and large timbers remained the only way to bridge wide spaces. For another, people tended to build barns the way their ancestors had done.

Halsted exhorted in 1881 that with the scarcity of heavy timbers, farmers should look at new methods of building, using dimensional lumber. "A first-class carpenter is not required," he wrote. Certainly not a first-class timber framer; and neither was a first-class barn, any more.

Only gradually did "pole barns" arise in the countryside. Aside from the decreased size of timbers, all sawn at a mill, the major differences were that the stud walls, not the posts and beams, bore the weight of the roof, and that the parts were put together with nails. No longer were special ways of measuring wood and cutting joints required. Although farmers were now able to be their own carpenters, they could no longer dismantle, move, and rebuild the same structure.

Few pole barns, even those that were maintained, lasted more than a gen-eration or two. Most of the barns deteriorating by the side of the road today are relatively recent, made of studs or at least sawn timbers. For new construction, wooden barns have given way first to metal roofs and walls, hung over metal trusses that are a kind of throwback to post and beam.

38

Dairying

The farm itself was changing, too. Farmers became far less motivated or able to keep up the old barns on their property. Changes in the dairy business affected barns as well as production. After dairy products were identified as a carrier for certain diseases in the mid-nineteenth century, states began to insist on testing. Pasteurization—heating the milk to destroy bacteria—was required from the first years of the twentieth century. Equipment for that operation and other new machinery didn't fit well in old barns. In spite of whitewashing the walls, it was hard to keep wooden walls clean, let alone a wooden floor.

By the 1960s, the federal government required farmers to have bulk tanks and concrete floors, instead of storing milk in cans and selling it from the farm or on individual milk routes. Insurance companies and banks didn't want cows kept in combustible barns. For most farmers it was easier to build a new, modern barn than to convert the old wooden one—as in fact it had been for 100 years or more.

Farmers in New England had to invest a lot of money they found it hard to get back, and not only for new dairying equipment. They also had to build roads so that tanker trucks could reach the bulk tanks. Since many of the old barns were remotely located, perhaps the only reason they were still around, they were simply too far away to be useful.

Because farmers were pushing their cows so hard—and pushing their crops hard to feed their cows—perversities of the weather could spell the end of a marginal operation. In a wet summer, farmers found they couldn't drive their old tractors through their fields to get their crops in, so they invested in larger, four-wheel-drive tractors. Their debts piled up.

Meanwhile, government price supports on milk, designed to keep farmers in business and thereby the nation fed, were keyed to what it cost to produce a hundred-weight of milk on farms much larger than those in New England and upstate New York. It cost more in the East, so some farmers found that they were losing money on the milk they produced. In 1996 Congress began to phase out a federal safety net for farmers, still recognizing some regional differences in milk production costs and approving the Northeast Interstate Compact, which may help New England dairymen and possibly those of New York.

Sometime toward the middle of the twentieth century, the development value of the land in New York–New England on which the farm was located surpassed the value of the crops that grew on it. At the same time, given the contemporary lifestyles of friends, it became harder for a young man or woman growing up to commit to the unrelenting labor of the keeper of animals. The majority of the Northeastern farms that were active in 1950 had sold out 40 years later to development or will be out of business by the end of the century, despite farmers' and governments' efforts to sustain them. Under these circumstances, it has been hard for the farmer to take care of those huge buildings called barns that no longer serve more than a marginal function, say, as a place to store equipment when it's not being used.

For that reason, barns built before the Revolution and still in use on working farms are virtually nonexistent. Most of the oldest barns Richard has found are well into the process of deterioration. All that holds them up are long-lasting slate or metal roofs and those massive timbers, hewn when trees were huge and plentiful, and pegged together by master builders with skill and pride.

III

GRAMP

RICHARD BABCOCK WAS BORN A GOODELL (pronounced with the stress on the first syllable). In the midst of the Depression, when he was an infant, his mother, Ruth Babcock Goodell, and his father, Clyde Goodell, broke up. They both went on to remarry, leaving the boy with a slew of relatives and half-relatives in upstate New York and western Massachusetts.

When Richard was a toddler, he first began to spend time on his maternal grandparents' four-acre farm on Cold Spring Road in Williamstown—hilly, largely rural country in the northwestern corner of Massachusetts. Clayton and Alice Babcock took him in when he was about one year old. He heard later that his grandparents had said, "If he's going to be around here all the time, then we want to adopt him."

Clayton Babcock would teach his adopted son the skills, nearly lost to history, of disassembling and reassembling timber-frame barns. Those skills, in turn, would become the key to understanding how and by whom the ancient barns were built.

That's how I came to be a Babcock, my grandparents' only son. I had the blood, anyways, from my mother, Ruth. She and my Aunt Isabel are the two survivors left out of four girls. The other two, Margaret and Edith, have passed on along with Gram and Gramp to that world where barns are always full of grain and hay, and the cows are full of rich milk.

As for me, I was a lucky son-of-a-gun, or so the saying goes. I only got a licking when I really deserved it, but when I did, I did. Gramp didn't fool around when it came to that and I'm glad he didn't. He had me so when he gave that certain yell, my ears came to attention and my body obeyed. Gramp used a leather strap to tan my hide. He would take hold of my wrist with the grip of a vice and I would do a dance as he applied the medicine I deserved.

Richard's maternal grandmother, Alice, was a Haley. Her father, George Haley, owned two farms in Williamstown, one of which was adjacent to the Babcock farm. Her nephew Richard Haley still maintains a small beef herd on the Cold Spring Road farm, which backs on meadows owned by the Sterling and Francine Clark Art Institute. Patrons of that elegant collection of American and French landscapes have thus been able to look out the window at real cows grazing on real grass.

Alice Haley Babcock's parents were George Haley, named after George Haley I, who had fought in the Civil War, and Rhoda Frazer, whose father, a Nova Scotia ship captain, had been lost at sea. Rhoda Frazer had migrated to North Adams to find work, meeting and marrying George Haley II. Richard remembers that Great-Grandmother Rhoda Haley, at her place just up the road, "always seemed to make the best of everything—bread, biscuits, pancakes, and plenty of butter and maple syrup to soak them down with. I mowed away some huge stacks."

His great-grandmother knew her Bible, having read it through three times,

family members claimed. "She would use it once in a while but didn't overdo it," says Richard. "She could play by ear on the vio-

Richard's grandparents, Clayton and Alice Haley Babcock, pose with her family, the Haleys. Alice's parents, George and Rhoda Haley, flank Alice in the front row; Clayton Babcock is behind Alice; her four brothers fill out the picture. (Photographer unknown)

lin anything you wanted to hear, beautifully old and mysterious sounds from Nova Scotia." Her husband, George Haley II, whom Richard remembers bucking up firewood next door, died when Richard was young.

Their daughter, Alice, "was raised with a tough group of brothers," Richard says. "My grandfather had to be tough to get a date with her. It was a match made in heaven for sure, because it lasted." Clayton Babcock dated her and married her.

He was the son of LeGrand Babcock, a storekeeper in North Petersburgh, New York, and Hattie Lewis. Richard has discovered that one of LeGrand's ancestors left hewn beams and tools in his will, an exciting suggestion that barns run generations deep in Richard's blood, but LeGrand himself was not skilled in that way. So when Clayton was a young man he apprenticed with "Square"

43

Allen of Petersburgh, a talented timber-frame carpenter who received his nickname because in his youth he was always playing with his father's carpenter's square. Thus Square must have learned timber framing from his own father who, in turn, must have apprenticed to a timber framer in his youth. That is the way the special knowledge was passed down, generation to generation, back to a family's arrival in this country and before that, to the old country. Richard has taught his own sons the skills, so that they can now trace the knowledge through at least five generations, over a century and a half.

When newly married Clayton Babcock came over the mountains from Petersburgh to Williamstown, one day early in this century, he stopped at a site

Richard's grandfather, Clayton Babcock, working on a bridge project in Adams, Mass., while in his 60s. (Photographer unknown)

where Williams College carpenters were trying to raise a structure made out of large timbers. According to the story that Richard remembers his grandfather telling, he watched as the boss carpenter on the job struggled to remeasure distances between posts that were rising skyward.

"I know how to save you some time," Clayton Babcock said.

The boss carpenter was miffed, asking who this stranger was. But the engineer said, "Let's hear him out." Clayton explained how to add a set amount each time, an old timber-framing method. When the system worked, the engineer hired Clayton.

Williamstown was a dairying town, yet it had one feature that distinguished it from hundreds of rural communities in New England. By some kind of crazy inspiration or act of faith, a college had been planted there, in the wilderness, in 1793. Most colleges in the newly independent United States had been founded along the relatively urbanized coast—William and Mary, Columbia, Yale, and Harvard. Dartmouth, in 1769, was the first college raised in the hinterland—Hanover, New Hampshire—established there specifically to attract

American Indian students.

What was Williams College's rationale? Only that a soldier in the French and Indian War left money to provide for the education of the children of a garrisoned fort in what is now North Adams. By the time the college was built, the war was over and the fort long-since abandoned. The college was founded, apparently, to teach farm boys.

The idea caught on, however. Dartmouth and Williams were followed a few decades later by a host of small colleges in rural settings. Those that succeeded became the economic mainstays of their communities, providing steady jobs that paid relatively well, thereby easing the dislocations as the mills have gone under and the farm economy continues to erode. Clayton Babcock was one of many carpenters and farming men and women to take a job with Williams.

The college also affected its community in indirect ways. The Williams art department was, in part, responsible for the decision of New York City residents Robert Sterling Clark and his wife, Francine, to build an institute for their collection of nineteenth-century French and American art in what was little more than a cow town in the 1950s.

Growing up with a college as a neighbor may have made it easier for Richard to commence book research. He first turned to Williams College's Chapin Rare Book Library when he began to study barn history seriously. Later research led him to registries of deeds and ultimately the Boston Public Library and the Harvard University libraries. Although Richard was far from growing up bookish, neither was he oblivious to what books could provide.

The Babcocks let the two cows they kept for milking make the rounds on George Haley's much more extensive fields. The Babcocks kept a vegetable garden and, by the age of ten, Richard kept his own small plot. Clayton Babcock stocked up on hay by picking it up on fields the college mowed, plus what he could scythe. "Gramp would do it for a while and I would get interested and give it a try," Richard says of scything. "That's how I learned everything from him. He would explain it as he would do it. If I wasn't interested in something, he'd wait until I was. He always seemed to know when the time was right."

Grandmother Babcock cooked up some great meals, like their own pork and greens—dandelions they dug up in the spring and topped with her own but-ter and vinegar. In the cold cellar off the basement she kept potatoes, eggs, sauerkraut, pickles, dandelions, tomatoes, and peaches, all in quart jars she had steamed on the stove.

Richard envied his Grandfather Babcock's grip, trying to acquire it him-self by milking the cows. In the afternoons he ran his newspaper route for the *North Adams Transcript,* hoping to beat his grandfather home in time to do the milking. He paid attention in school, he says, but could not find time for the homework afterwards. Some of his classmates still in Williamstown are sur-prised that Richard grew up to be an expert on anything academic, such as barn history.

He played sports in high school, serving as captain of the gymnastics team under Coach Ted Sylvester. "He knew how to teach," Richard says. "He could do it himself, and you got interested watching him. Then he'd praise you when you did it right." In small Williamstown High School, Coach Sylvester's all-star football team played the same 11 men, including Richard, on offense and de-fense, with only one substitute. Coach called it his "iron man team."

Occasionally on weekends or in the summer Gramp would have me along on small jobs doing a little nailing, something ev-ery boy likes to do. Even that was an art to Gramp. Every nail had so many hits to put it in. The way you held your hammer made the difference. Some I've seen today, supposed to be carpenters, don't know how to hold a hammer.

He showed me how to saw, as well, to eye over it so as not to see either side of it, and to go steady and straight with the push and pull of it. "Let the saw do the cutting," he'd say.

He would take me out on framing jobs he and Charley Haley would do together. Charley was old George III's son, my cousin. Everyone admired Charley, who was always quiet and sure. He was

46

a hero to me, too. He and his brother, George Haley IV, fought through the Second World War. Charley and Gramp kept me nailing steady on these small jobs out in the sun, and I loved it. Charley and Gramp were the genuine article when it came to carpentry. They did it good and fast.

Richard worked summers on farms in the neighborhood, including the Haley farm, with his cousins Richard and Robert. Their older brother, Jimmy Haley, did the plowing and mowing with a team of horses. Jimmy learned to farm the old way, and Richard appreciates that he, too, learned from Gramp and Old George how to plow with horses.

"There was always more hay to bring in," he remembers, "and we loaded it by hand with a fork, loose. I loved the sun even through the dust and the hayseed"—and the horsefly bites. "My arms and legs grew strong, and I was proud of my frame. Gramp was, too. He'd praise me up once in a while." Once Richard was riding on top of the pile on the fully loaded hay wagon when a wheel hit a stone while crossing Hemlock Brook. As he bounced off he realized the hay was coming after him, so he pulled himself under the wagon—fortunately, or he would have been smothered by the hay or drowned in the brook under the load.

Another time he did not escape unscathed. The middle finger on his left hand was crushed by a sled loaded with two cords of firewood. "It was cold," he says. He was out alone, when the sled's runner passed over his hand. "I tucked my hand under my arm and drove the team home. I had to rest twice on the way. When I got to the house, Gramp came out and unharnessed the horses. My God, it was cold." Now he has a space, less than a full finger.

When World War II came, brothers Charley and young George enlisted, leaving Jimmy to do all the farm work. Robert took agriculture courses at Williamstown High School, so when Jimmy left plowing with a team to become a carpenter and old George bought his first John Deere tractor, Robert had learned how to care for it.

Enjoying being outdoors in the sun more than studying, Richard left school in 10th grade to help install a telephone line across the hillsides. "Walking in each day and digging post and anchor holes was hard work and brought me to age eighteen," Richard reports. He and a friend, Ron Leonard, joined the Marines together during the Korean War, getting kicked into shape at Parris Island, South Carolina. Although Richard wanted to be a rifleman on the front line, at first he was sent to Quantico to skirmish against those being trained as officers.

He was halfway across the Pacific on a troopship when truce was called. His ship was rerouted to Japan, where the men trained in the mud of a monsoon to be ready in case the truce broke down. After eighteen months of mingling enviously with the men who had actually fought, Richard, the acting platoon sergeant of the Third Marines, returned to the United States. He married Sally Noyes, a Williamstown beauty, and they moved to Cherry Point, North Carolina, until his enlistment was up.

His wife was pregnant, living off base, when a hurricane ripped through. Unable to reach her by telephone, Richard made his way to the house only to find it entirely cut off by a risen creek. He dove into the swollen river and swam across. The heroic swim was for nought. She had been evacuated.

After they arrived back in Williamstown in October 1955, their son David was born in nearby North Adams Regional Hospital. Richard remembers that the Taconics, the Green Mountains of nearby Vermont, and even Mount Greylock, the tallest in Massachusetts at 3,491 feet, looked like bumps compared to 12,395-foot Mount Fujiyama, which he had climbed when he was in Japan. He had watched the sun rise over the Pacific from the summit while beneath him the rice paddies looked like a checkerboard. But now that he had seen something of the world, he was ready to settle down.

Richard bought a 225-acre farm in Hancock, about 100 acres of it tillable, the rest on a mountain. Gramp knew his adoptive son liked farming. Before he had gone overseas, Richard had worked at the Haley place and earned two calves from their "exceptional" cows. While he was away his grandfather had turned them into seven.

Richard teamed up with his grandfather, working for Williamstown contractor Davy Dean. "I joined the carpenter's union with Gramp and started at $1.25 an hour," he says. The jobs Richard got didn't involve real carpentry, however, and he was itchy to get to farming. One day he told his grandfather he needed a barn bigger than the small one, just a shed really, on the farm he had purchased.

Gramp said, "Why don't we move one?"

Richard replied, "That sounds good. How?" They both left their jobs, while Gramp showed Richard how. "Those were the greatest days, just he and I taking down a barn."

Richard as a U.S. Marine, during the Korean War. (Photographer unknown)

Gramp told Richard of a barn on his father's old farm in Petersburgh, owned then by his brother, Class Babcock. They went over to check out the building. Clayton yelled slowly at his deaf brother, "Could we move your barn?" And Class bellowed back, just as slowly, "Sure, take the barn, it's only going to go down, otherwise."

It was grown up around the 30- by 40-foot barn. Some of the stable windows were out—otherwise it stood up straight, painted red with a wood shingle roof. The doors were all stuck, but we forced one open and started looking around. There was some old hay in the mows — pretty brown — probably twenty years old. There were some pigeons living there that entered by the vent holes in the gable ends. There was an old bean thrasher, Gramp called it. He told me how it worked. Showed me a little. He started using words like purlin plates, cross beams, girts, and such, talking about the structure of the barn.

I asked, "How do we get it down?"

He said, "We take off the roof first — then take down the rafters. Those are what support the roof boards and shingles."

I judged the barn was old — it looked old. According to Gramp, his dad kept cows and horses there. Gramp said we'd need a gin pole tree, a block and tackle — that's pulleys and ropes — planks to walk on, and tools. We came back later with Gramp's three-quarter-ton pickup to take her down. Oh, Gramp brought a ladder, without which we would have had a lot of fun, and a lunch of cheese and crackers.

When we sat down for a break he would tell me how the old timber framers felt about the modern buildings. He told me old Square Allen said, "It's all over now. Nobody cares any more about the old framing methods. All they want to do is sock the lumber full of nails."

IV

GIN POLE AND BULL WHEEL

BUILDING POST-AND-BEAM BARNS requires skill in the arcane art of timber framing. Even just the raising and lowering of these structures requires special knowledge. Enormously heavy beams must be heaved into the air—or taken down, if a barn is to be moved and rebuilt. A related issue is how to add connecting members that must be slid into mortise and tenon joints; and how to take them apart. Furthermore, most of this must be accomplished by people who are standing on the ground most of the time.

In putting up and taking down barns, the old way is the best way. The answer was not and still is not jacks, backhoes, or even motorized cranes like one Richard designed. "I have to use the crane sometimes, because it's cheaper than paying people," Richard says. He never uses it on barns with a special history or where the raising requires authenticity.

The best way is to use a gin, a device related to the cargo hoist in a freighter or a ship's rigging. The word "gin" is thought to be an abbreviation of the word "engine" which, in turn, relates to "genius" and "ingenuity." The Middle English word *engin* referred to native talent. "Gin" therefore derives from words relating to the clever solution to a problem.

Derrick-like structures for lifting weight go back to prehistory. Often these were in the form of a tripod, similar to the contraptions country kids still build to haul the engines out of their automobiles.

A gin pole substitutes three guy lines for two of the legs of the tripod, so that the device can be more easily moved around the site. This three-point hitch is similar to the stays on the mast of ship.

A block, made up of one or more pulleys, is attached to the upper end of the pole, which in turn is positioned over the timber to be raised. The lower block is attached to the timber. A rope passes over the pulleys, each turn providing more leverage. The pole can be leaned slightly, but it loses strength as soon as it leaves the vertical. Richard has experimented with a jib pole, mortised into the gin pole on an angle, for added maneuverability in lifting lighter beams such as rafters.

The bull wheel, related to the capstan on a ship, can help take up the slack. It is a drum, fastened to something solid, that can be rotated horizontally by a pole inserted through the central shaft. The bull wheel can be turned by animals, as the name suggests, or by human beings. Richard rediscovered the device in the writings of the eighteenth century French encyclopedist Diderot, rather than learning this from Clayton Babcock. The use of nautical devices in barn construction reinforces the historic relationship between barn building and ship building.

The bull wheel

Richard's grandfather taught him how to raise a barn by first taking down his own father's two-story one.

52

We went up the ladder to the eaves—the bottom edge of the roof itself. Gramp wasn't a bit afraid of height. He started off the ladder at the base of the roof—leaned forward a bit and walked, feet flat, right up the roof. Although I was scared of sliding off, I followed.

Starting at the peak, we used our bars and hammers to take up the roof boards and with them the wooden shingles, as down we came, gripping only with our feet, grabbing onto the board we were pulling up. Sometimes we had to pull a bunch of shingles free to get to the boards. We found birch bark used to patch the roof, slid up under the shingles and looking as good as new. I also saw metal can lids used as patches.

Swinging the boards around on end, we let them slide down the roof. They hit the ground on their ends and then fell back against the side of the barn without breaking. It worked well on the side of the barn built against the hill but not so well on the lower side. On that side we passed the boards through a hole in the roof, dropping them on end into the old hay. When we had a bunch of them down, we'd go down and stack them outside.

There were some bees, hornets, and mud wasps around our heads, not too happy about what we were doing. Gramp said, "Don't pay attention to 'em and they won't sting you." Well, that was a little hard to believe the way they'd dive-bomb me from time to time.

I got a little scared, but Gramp didn't pay them any mind, so I tried to act like I wasn't scared, either. It seemed to work, but somewhere along the way they figured me out and I got stung. Gramp didn't. Still, I learned to deal with them without spraying. I'm glad

I learned from Gramp.

Once the roof boards were off, we knocked the wooden pins out that held the rafters and slid them down the roof on their tails, leaning them back against the barn. This we did with more care than with the roof boards.

Soon we were walking on the purlin plates, beams that run under the rafters about halfway up, or standing on the outside wall plate. Plates are horizontal members, like the sills but overhead. Grampa walked the plates as though he were strolling on Main Street, with a two-story drop below, kicking off bird manure as he went.

When we sat down for a moment in the course of things, he told stories about old barns and old barn builders. Somewhere in there

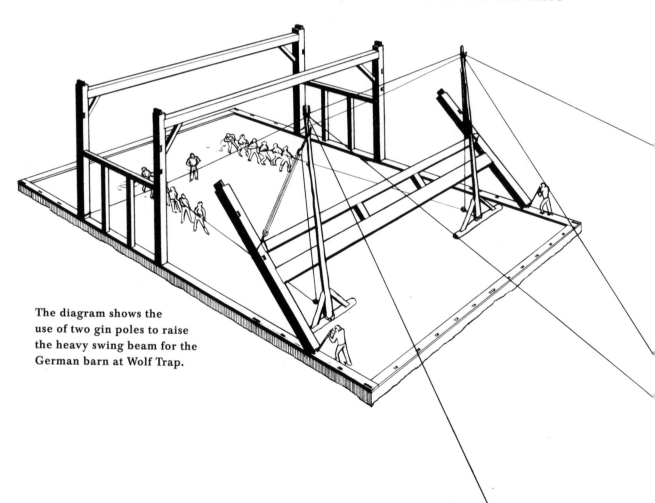

The diagram shows the use of two gin poles to raise the heavy swing beam for the German barn at Wolf Trap.

I asked him, "How do we get the big beams down?"

An ancient barn once belonging to the Galusha family is disassembled in Williamstown to become a home. The view is of the Hopper on Mount Greylock.

He replied, "With a gin pole and rope." He explained that we would have to cut a gin pole from a tree. Later he drove the old pickup over a forgotten road across a field and cut the barbed wire fence to let us into an old, upland woodlot. He said a straight maple or beech would do. This time he cut an ash. "You find a good branch," he said, "then measure down from there to get the height, about six feet above the highest point you have to lift. The branch holds the ropes when you stand her up."

He cut it off where he figured, and we used the truck to drag it to the barn. Using a short piece of log as a roller, we slid the pole into the barn.

Gramp had his old pulleys and ropes—block and tackle he called them. He strung the rope through the blocks, back and forth, showing me how to do it. When he was done, he tied three long ropes to the top of the pole, the guy lines he called them. To pull up the pole, he fastened one block to the purlin and the other block he fastened temporarily three-quarters of the way up the pole. We took up the slack, hauling it up. When it was standing we were able to move the butt perpendicular.

He directed me to tie off the guy lines, about equally apart, to beams, the truck, or anything solid. I was learning how to rig a gin pole. Standing on the purlin, Gramp moved the block up to the branch and fastened one loose end of a line to the block and tackle. Using an iron pin about the same size as the wooden ones, we hammered them out at each post. We hauled on the line to lift the purlin off its posts and lower it to the barn floor. With rollers, we moved it to the back of the pickup.

Gramp sure had a bag of tricks, balancing, sliding, rolling; once in a while a bit of hard lifting. He used a lot of leverage.

We moved the base of the gin pole and adjusted the guy line to keep it perpendicular until it was next to the other purlin. Then we fastened the block to the beam, hammered out the pegs, hoisted that purlin up, and lowered it.

Gramp's Toolbox

Clayton Babcock used hand tools, not Skilsaws or Black & Decker electric drills. He didn't have to plug anything in. He didn't even use an adze or a broad, hewing axe. He hewed the beams with a small, cutting axe, sharp as a knife. He used a hammer, pry bars, and a steel rod to drive out the pins that fastened the joints. Sometimes the pins were driven into holes deliberately offset, in order to make the fit snug. They could not be hammered out.

Gramp told Richard to drill the pins out, using a brace and bit. That wasn't easy, Richard says, because sitting astride a plate a story or two in the air, he had to pull on a beam with one hand while pressing on the round handle of the bit brace with his chest, at the same time turning the handle of the brace with his other hand. The wooden knob on the end of the brace has left a permanent dent in his chest. "End grain is hard to cut," his grandfather told him. The only way to ease the job was to keep the bit sharp, which Gramp accomplished by applying a three-corner file. Richard remembers a dozen oak pins that had to be drilled out on Class Babcock's barn.

Gramp used files or a Carborundum stone to keep the tools sharp. Sometimes he simply reached down to the ground and picked up a stone—not too hard and not too soft—to touch up the edges of the ax or chisel.

Gramp used a big hammer he called the "gentle persuader." The head was made out of the trunk of a small tree, bound with two metal straps. He used a crosscut saw and other handsaws. He had a couple dozen chisels with which to make the mortises and tenons. He kept a big slick, a blade pushed by the hands (never struck with a hammer), to smoothe a tenon.

Mortises are the slots into which the tongues or tenons fit. These joints can become complex when three or more beams come together. Tenons usually run into a slot that goes partway into the beam. Generally a peg is driven through the outer edges of the mortise and through the tenon to secure it, although in the Dutch barns the tenon pro-

The mortise and tenon (shouldered version)

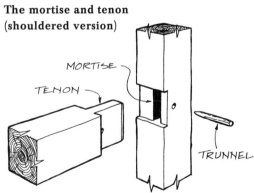

MORTISE

TENON

TRUNNEL

57

trudes and can be wedged on the outside of the post. Joinery—Richard uses the earlier word, jointery—is the art of making these connections.

Gramp used a bubble level. He told Richard that in the old days, builders found the level with a plumb bob mounted in a triangle. When the point on the bottom of the bob or weight hung over a point directly beneath the apex of the triangle, the bottom was level.

He used a chalk line and a builders' square. He seldom used his old carpenter's compass, a set of dividers with iron pins about a foot long, joined at the top, which can be set at different distances apart. But he taught Richard how to use the compass and how to apply the "scribe" or "square rule."

Right Angles

Today builders use metal squares made in factories to make right angles. They can be placed against the straight edges of dimensional lumber. Old-time carpenters had squares, but they didn't work the same way on irregular, hand-hewed timbers. The squares were useful along the charcoal line to mark parallel lines—two sides of a mortise, say.

In the old days, a man used a carpenter's compass, its legs fixed a set distance apart by tightening a screw or adjusting a wedge. To make a straight line, he rubbed a string with charcoal and then drew it tight between two points, pulled it, and let it snap back, leaving a black line. Today such a line is rubbed with blue chalk. The color makes no difference; the idea is to pull it tight on the wood between two points and snap it, leaving a straight line on the surface.

The old builder (and now Richard, to demonstrate the old ways) used a compass to set the lines to hew. He flattened an area at each end of the log and inscribed two circles, snapping the charcoal line between their perimeters near the outer edge of the log. The builder set the compass to whatever setting was appropriate and walked it down the black line, first one foot of the compass on the wood and then the other. Thus distances were measured in numbers of turns of the compass. Richard believes these circles, moved outside, are the origin of hex marks—the builder in his pride leaving his mark.

The old builder used combinations of turns for the various members, and these combinations, reduced to a series of circles, became his master settings. He transcribed them on a board, which he used as his building plan. Occasionally Richard has found these boards on the barns he has dismantled. "By re-establishing the lines and setting my compass to his settings, I came out right on his points," says Richard. He underscores that this way of measuring was accurate for timber framing.

Richard doesn't think much about inches or feet when he's framing. Like his predecessors, he thinks about the number of turns of the compass.

"At each intersection where girts, braces, and beams joined," Richard continues, "there was a certain turn of the compass, so that each section of timbers would join together when the structure was raised." The pieces that came together at each major joint were marked with pairs of Roman numbers, such as I, II, III, IV, which were easier to strike with a chisel than curved Arabic numbers.

The Scribe Rule

The scribe rule, a secret revealed only to the most worthy of a master builder's apprentices, enabled the construction of joints, framed on the ground, to prove square when the timbers were raised. Gramp showed Richard about it but they did not use it on the re-assemblings, so Richard had to rediscover it later, when he was grappling with forming square corners out of irregular wood.

Jack A. Sobon, a fellow timber framer who worked with Richard, showed him a passage in Edward Shaw's *Civil Architecture* (1832?) that Richard believes misidentified the scribe rule as a method of etching the correct lines by laying one timber over another. The book said that was the way of framers who didn't know how to square a building properly. It required moving heavy beams.

Shaw describes scribing as follows: "For the tenons you are about to strike, place the lower edges of the girder to the line of the lower end of the mortises; make a scratch on the girder at both ends, exactly to the face of the mortise." Shaw writes of what he calls the square rule: "This method has the preference in detached framing: the timber admitting of being framed in different places, and not tied together until its raising."

Later Richard put together what his grandfather had told him with the

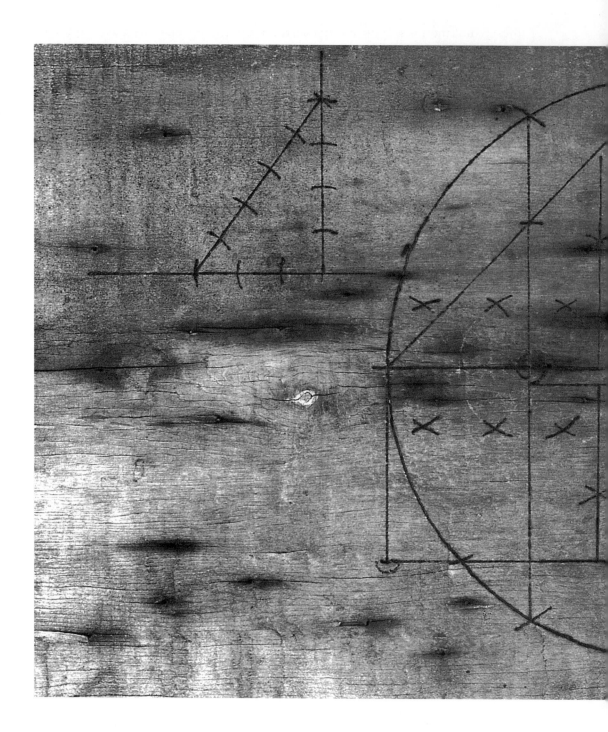

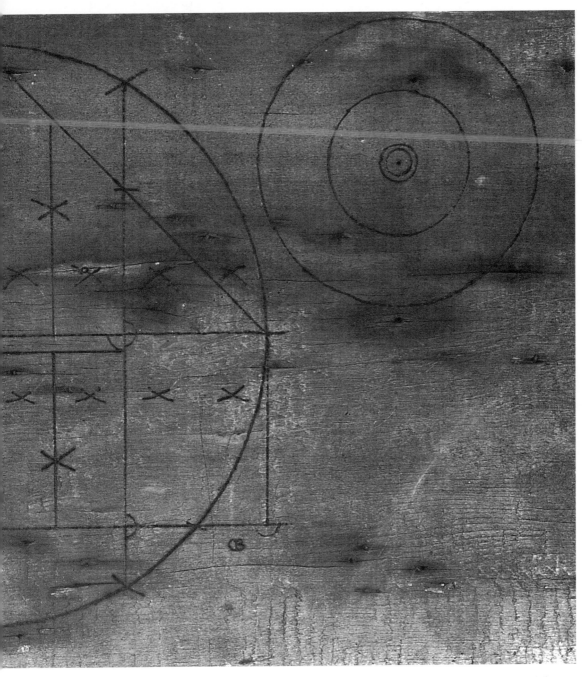

Richard found this complete plan for a German or Dutch barn, center, drawn by a master builder. The three-four-five triangle, upper left, shows how he squared his sections; the concentric circles, upper right, are the setting for his carpenter's compass.

traditional practice of using the Pythagorean theorem to square corners of a foundation. That method requires inserting pins in a line to form a three-four-five right triangle. It occurred to Richard that he could use the compass turns to do the same thing. The trick was to imagine and then mark the perfect timber that the irregular one contained, by snapping a line between the centers of the circles and using that, rather than the edge of the beam, for measuring.

Builders carried with them small pieces of wood they used to form the hypotenuse of the triangle. With the small piece of wood as hypotenuse, they adjusted the two beams until five turns in the hypotenuse met three turns on one beam and four turns on the other. Richard knows builders used these, because his grandfather had one.

The scribe rule was not just the way of getting by for the uneducated. Richard had rediscovered the real scribe rule, not how to transfer lines but how to make sure that joints would be at right angles before trying them out. Unlike what some books said, Richard believes that the scribe rule and the old square rule are one in the same.

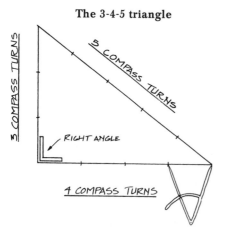

The 3-4-5 triangle

"Gramp told me the old masters wouldn't lay out the frame when anyone was watching them," Richard says. Using the three-four-five triangle was one of the secrets of the trade, zealously guarded.

Sobon became intrigued by the two ways of laying out timber framing, by whatever name they were called. While agreeing that the old timers knew how to square using a line and compass, he still thinks that the term "scribe rule" referred to tracing or scribing locations for joints by laying timbers on top of each other on the ground. He notes that it was a highly developed technique, as cathedrals were built that way.

He believes the "square rule" originated in this country, first because straighter timbers were available and then because wood began to be cut with straight edges, in mills. Richard believes the square rule Sobon refers to is "hidden within the scribe rule."

Regardless of the terminology, measuring and squaring with a compass and line enabled the builder to lay out frames more conveniently than tracing did. The "square rule" saves about half the time of the tracing approach, with its lengthy preassembly procedures. To understand the improvement, we must remember that the timbers involved were often so heavy that it took several men just to turn them over.

On a break, Gramp told Richard about a man and his son who used a horse to take down and put up barns. The horse was an old one, who wouldn't do something unpredictable that might bring down a heavy beam on someone's head. They used the horse to skid a gin pole out of the woods and to take up the slack on the block and tackle. Having learned how to work with horses as a boy, the idea intrigued Richard; yet he has never had the opportunity to use a horse in his barn work.

Richard and Gramp put temporary braces on the wall posts so that they could remove the siding. They moved the salvageable wood in several truck loads to Richard's place. Some of the beams were totally rotted and some had small rotted sections. Gramp showed Richard how to splice in good wood and how to hew out replacement beams. In his career Richard hews mostly for demonstrations or other cases where only hewn replacements will do. For commercial purposes he is more likely to use beams from barns he keeps for spare parts.

The sills of Class's barn were mostly gone, something you couldn't see easily when the barn stood up with a floor on it, especially when there was hay in it. Gramp asked me if I wanted to use new, sawn timbers for replacement or try some hewing. I said I wanted to hew some, without knowing what I was getting into. He said we'd have to find the right trees. I suggested we cut in 30 wooded acres of my property along Hancock Brook, through which a logging road ran.

Gramp walked to an old oak, looking up it as he circled it. He said we should be able to get the 26 feet out of it we needed to re-place the cross beam that had rotted underneath the upper hay door. That door had been missing from Class's barn and the rain had gotten in. In the same way the front sill was rotted underneath the main doors, which must have been left open in the rain too often. We found a small oak for the sill and shorter trees to replace some girts. Gramp took one end of the crosscut saw and I took the other.

We cut the logs and we cut two 4-foot lengths on which to place the beam we were hewing. Gramp had me help him roll the log around a bit while he eyed down it for the straight way. I drove the iron dogs—overgrown staples—into the log to be hewn and the short one below, at each end. Then he used his ax to flatten two spots directly above the short logs. We removed the dogs and rolled the log 180 degrees. After we replaced the dogs, the log was stable. Then we flattened spots on top as wide as we needed the width of the beam. Gramp said we had to get below the sapwood. He made a circle at each end and snapped the chalk line on each side of the log.

Standing on the log, he drove his ax in hard to reach the chalk line on one side. Then he swung toward that cut two or three times, at a 30-degree angle, until a big chip came out. He told me some of the old hewers would spit tobacco juice down the sides of the log to plumb the cut. "If it doesn't stain all the way down, you're undercutting." Gramp didn't chew; neither did I.

"Practice makes perfect," he said, handing me the ax. Then he said, "You shouldn't have any trouble sleeping tonight." Standing

Richard hews a gunstock post for the Animal Rescue League barn in Dedham, Mass. (Harding-Glidden photo) ➤

64

on the log took some getting used to. Driving the ax in to the line was a trick. We took turns, he driving in at that 30-degree angle, scoring it, then chopping it off clean. I struggled even to keep my footing. It was hard to get used to eyeing down the side, to get the edge plumb.

Once one side was done, checking it with a plumb bob, we rolled the log to do another side, until all four were square. It was easier to stand on a flat surface. "The first side is the hard side," Gramp said. "After that it's all downhill." It took us a week or two to finish our beams.

Then we put them aside to cure for at least a month while we laid up a foundation of stone, squaring it in the old way forming a three-four-five right triangle with string. We rolled in our floor beams, some old, some new, always putting the crown side up. That is, any bow in the wood should be on top. The weight of the structure and the hay will flatten it.

"It was better to work with chestnut in the old days," before the blight wiped out the species, Gramp said. "It was a great wood to work with, it was resistant to rot and you never got tired of eating the chestnuts."

Then we laid new plank flooring over the beams. Setting up the gin pole on the deck was like raising the flagpole on Iwo Jima. He held down one end and I walked up underneath the other, lifting it as I went. That was where my strength came in handy. We had already tied on the block and tackle and the guy lines.

We raised the posts the same way as the gin pole. I held them up while Gramp nailed in a brace. The first post we fixed leaning

out about a foot. Then we rolled the
end cross timber in front of the posts.
We centered the gin pole and raised the
cross beam using the block and tackle

Richard turns the bull wheel for extra
leverage in raising a bent at Philipsburg
Manor. For that job the crew dressed in
18th-century garb. The line from the bull
wheel runs through the block and tackle
that hangs from the gin pole.

until we were able to slide its tenon into the mortise of the post that
was plumb. It required some finagling and pounding on the back of
the post. Then we lowered the cross beam over a center post.

"How do we get the leaning post over the other end?" I asked.

He said, "Stand your ladder from the ground on the outside
against the top of the post. Climb up and drive it in with the
persuader."

When I got up there he released one of the braces so the post
would come in. It did just that when I hit the post against which
my weight was leaning.

67

So it went until just the two of them had erected a medium-sized barn. As mentioned, Richard believes that the picturesque barn raisings, when all the neighbors arrived to help push up assemblages of beams and plates called bents, were few and geographically far between. If a master builder and an assistant followed the old ways of applying leverage, the two of them could accomplish most jobs.

Later Richard first read about and then found the remains of a bull wheel in a barn in Schoharie, New York. Richard says that because Gramp had taught him the fundamentals, he knew immediately how the bull wheel fit in when he read about it. It was a way of applying even more leverage while taking up the slack. With its aid a handful of people could put up even the largest and heaviest oak bent.

Richard found a drawing of a bull wheel, which had many applications outside of timber framing, in the French encyclopedist Denis Diderot's writings. He and his employee Henry would later spend a week reconstructing one he found for use in the barn raising at Philipsburg Manor, Tarrytown, New York.

A New Direction

While he and his grandfather were raising the beams with a gin pole in Hancock, passing motorists frequently stopped. One man stopped every day, even after they had begun to close in the building. As they were nailing on the last of the siding, he started down the dirt road to the barn. "Well," Gramp said, "I guess we're going to find out what's on his mind."

He introduced himself as Joe Starobin. He asked where they learned how to put up a barn. Richard pointed to Gramp, then in his sixties, and explained that he had learned how to do it long ago. Starobin said he was thinking of putting a barn on his nearby place and converting it into a ski lodge. He said he understood there was a barn in town that was in the way of a road. He was going down to New York City for a couple of weeks. He wondered if they could give him an estimate on taking it down, moving it to his place, and putting it up.

The next Monday Richard found he could have the barn for a dollar, but the catch was it had to be out of the way in two weeks. Richard made a beeline to tell Gramp, who asked what would happen if Starobin didn't want it. Richard answered that he'd keep it himself. Gramp said, "All right," that's all, and the two of them were at work before the day was over.

This barn had a slate roof. They put up pole staging along one side. Gramp showed Richard how to lift the slates with a hammer, leaving the nails in to slow them down as they slid down the roof. Gramp caught them at the bottom, piling them on the staging until it wouldn't hold any more and they had to move them to the truck.

Each day as they worked, the bulldozers drew closer. After two weeks the little car pulled up. "I didn't tell you to go ahead," Starobin said, alarmed. He calmed when Richard explained he would take it himself if Starobin didn't want it. He also explained total cost to date: One dollar for the barn and $600 for two weeks' labor for two men. Starobin was satisfied when he heard the two of them would continue working in between milking the cows and other farm chores. Richard hired half-brother Fred West to speed up progress.

In June of 1963, while they were raising the first beams, Starobin brought over a reporter and a photographer from *The Berkshire Eagle*. The resulting article drew a call from retired General Horace Harding of Lenox, who said he would give Richard his old barn if he would take it away, because the stone foundations were caving in. As Richard started work on that project, Harding asked him what he was going to do with the barn, explaining that he had a friend in Mill River, Massachusetts, who wanted to make a home out of it. Gramp stuck with Richard on that summer home for Richard and Katherine Cunningham, helping build Richard's first stone fireplace.

Richard forgot about trying to make a living out of farming and has been moving barns ever since.

V

DISCOVERING
HISTORY

WHEN GENERAL HARDING ASKED Richard about rebuilding the barn as a house, Gramp said, "Why not?" He knew plenty of people who had had to live in their barns—usually after their house had burned or before they could afford to build a house. What the general had in mind was different, however. The Cunninghams wanted to live in a barn because they thought they would prefer it.

Later Gramp explained that he understood that, too. People had converted barns to houses long ago, covering up the beams so that they didn't appear to be barns.

"Why not let them show?" Richard asked.

"Where are you going to put the insulation, wiring, and all?" Gramp countered.

Richard suggested re-erecting the frame and nailing the old siding back on, where it was, visible from the inside. As the next step, Richard proposed adding a stud wall around the outside, built up from an enlarged foundation. The utilities and insulation would go between the old wall and the new. The new siding would leave the building weather tight, while the old would give the inside character.

The Cunninghams thought that was a great idea. And it could be said that the new wave of barn conversions and an enormous increase in timber-frame home construction in the last twenty years both owe a great deal to that conversation.

The newspaper publicity and word of mouth led to a string of barn movings and conversions, and Richard drove about the countryside developing an inventory of post-and-beam barns. Other contractors picked up some of Richard's ideas, after having heard or read about them.

Gramp was then in his seventies and, although he helped out from time to time afterwards, the Cunninghams' was the last barn he worked on full-time. Not that he retired; rather, he returned to working for contractor Davy Dean. He was beginning to lose his eyesight, so that when people came up to him he would have to listen for a while before saying, "Now I know who you are." Other than that, most people didn't realize his vision was failing. He knew his way around carpentry jobs so well, he could pretty much learn what he needed to by feel. Richard remembers him asking on one job whether what he was touching was a nail or a knot, however.

Williamstown resident Michael Steele reports that Gramp was on the crew when Davy Dean was installing machinery for the manufacture of photographic film in the old Williamstown Manufacturing Company mill on Cole Avenue. Quite a bit of carpentry was involved, building platforms and providing access to the equipment. Visiting the site one day, Steele could hear the sound of hammering way back in an unlighted corner of the rambling mill. Sure enough, there was Gramp, working away. It didn't bother him that it was dark.

Right-Hand Man

Richard could not carry on by himself, however. One day, while he was driving to the Cunningham project, he saw a man striding down Route 22 near Stephentown, New York, a man who "looked as though he were bound for work." Richard offered the man a ride, looking him over carefully. "His hands were calloused and there wasn't an ounce of fat on him," Richard later wrote.

"His clothes were worn by work. Beads of sweat rolled down his cheek and fore-head." Knowing a good worker when he saw one, Richard offered the man a job. His name was Henry Suyden, and he has been Richard's right hand for forty years.

The son of rural wood choppers, Henry lived with his family in an isolated one-room home in the forest 1.75 miles off the paved road. The family burned kerosene for light and pumped water. Henry did some carpentry, farming, and wood chopping. He says his father's name is French; his mother was a Packard from New Ashford.

Henry's cars were none too reliable, so Richard would swing by to give him a lift. Henry remembers when they picked up bottles on the roadside to cash in the deposits to get gas, and when they pretty near starved waiting for a client to pay.

"There were good times and bad," he says. "Sometimes we didn't have two nickels." Indeed, Henry papered one wall of his kitchen with Richard's bad checks. "Sometimes," he says, "there was only the two of us to lift 50-foot beams." Henry remembers Richard's old farm truck, so loaded up with timbers that the only way into the cab was through the broken rear window.

Sometimes Henry got mad and stormed off. "I always went and fetched him back," says Richard. "Fortunately for me, he always wanted to come back." Over the years the two men became tolerant of each other.

"I enjoyed it," Henry confesses. "It was a different kind of work. Every barn was different."

Could anyone less accustomed to unrelenting work have stayed with Richard so long? The ex-Marine frequently swore a blue streak at his employees. He didn't tolerate loafing. He kept his own counsel. As one former employee, Paul Martin, says, "Life with Richard was an adventure."

Sometimes the adventure was humorous, at least in retrospect. Martin remembers framing the Davidson house in the field behind Richard's home on a hot summer day, prior to moving the beams to the site and erecting them. He had left the side doors of his van open. When he

Henry Suyden, Richard's ➤ right-hand man for 40 years, is hoisted up a gin pole to add another line for some heavy lifting at Wolf Trap.

returned to it, he couldn't see inside, which seemed strange to him. Getting closer, he saw what had happened. One of Richard's cows, seeking the shade, had climbed inside the vehicle, pretty much filling and trashing the interior.

What Henry Suyden and Paul Martin remember even more vividly is the unpredictability of their boss, which often resulted in the crew members working on their own. Someone once asked Henry where he was going to be working on a particular day. Henry said, "I don't know until I get in the truck. And sometimes even then he swings around and heads another way." Richard is known for changing his mind at the last minute and for making side trips that leave his employees guessing at his whereabouts.

Henry was in charge of breaking in new employees. "If they could keep up with Henry for a day," Richard says, "or at least if they tried and didn't give up," he could be pretty sure they would be a help. In Richard's opinion, Henry "did a day's work."

Paul Martin and Jack A. Sobon, another employee at about the same time, tried to bring some order to the apparent chaos, sharpening tools and putting on new handles. Sobon even built Richard a toolbox, so that hammers and chisels weren't always falling out when someone opened the truck door.

Both Sobon and Martin say they owe Richard a tremendous amount and both respect him. They both went into timber-frame construction on their own after working with Richard, Martin working out of Bennington, Vermont, and Sobon out of Windsor, Massachusetts.

Learning More

In the early 1960s, Richard started using rainy days or days between jobs to research as best he could the histories of barns he was moving or had moved. He matched what he learned through books and records to what he saw with his own eyes.

In one case, the barn that became a ski lodge was originally built beside the road, just across from the cemetery where the probable original owners of the land, Asa Clothier and his family, are buried. Clothier, who fought in the American Revolution, and his family have handsome stones in that cemetery.

Richard believes that nearby scattered, unengraved stones mark the resting place of the family's slaves.

Richard's son, Clayton, pounds in a peg while his father supports the girt. The photograph, taken at Philipsburg Manor, shows how good wood can be pieced in to a bad section of an anchor beam.

The barn was 26 feet by 36 feet, by 24 feet to the peak. The wood was mostly beech, the timbers chopped out by an ax and framed by the scribe rule. It was not one of the early barns that show their European roots; rather it descended from an English type. After a Clothier married a Jenks, the barn remained in the Jenks family for a number of generations. When Richard bought it for one dollar, the owner was Edgar Whitman. The barn now stands 50 feet higher than it did and a quarter mile away, and serves as a rental home.

Richard moved two more barns to his own farm and converted them, with the idea of setting up a kind of barn-house village. Although he was able to sell them and the farmland they were on, the idea was not a viable one. All in all he has moved and converted six barns in Hancock, while gradually extending his net to other towns in western Massachusetts and eastern New York State. He says he watched hundreds of barns rot into the ground—barns that the farmers couldn't afford to fix up and that Richard didn't have clients, money, or time to rescue. The number that he could save was necessarily small in comparison.

Changes

Richard's first marriage ended in 1959. "Sally and I broke up, through no fault of hers, after having five children," he says. "I thought it was her fault at the time. If I had known the strength of love as well as I did the strength of timber,

75

we might have made it." The children—David (born in 1956), Hannah (1957), the twins Alice and Susan (1958), and Charlene, born in 1960 after the breakup —left Richard's life to live with their mother. He saw them rarely while they were growing up, although he thought about them.

Not long after the breakup, Richard met a waitress at a restaurant he frequented in North Adams. Donna, whom he describes as the answer to his prayers, and he were soon married. They also had five children—Charley (1961), Clayton (1963), George (1964), Ronald (1965) and Margaret (1967).

Getting the Lead Out

Richard worked hard. He was still tending his farm, just as his forebears had done regardless of their main source of income. He maintained a small dairy herd and drove his old tractor out to mow, replacing the muffler when tree branches knocked it off. When he moved his barns, he often had to clear their new sites, cranking up the chain saw to cut down the trees, although he never used a chain saw on his barns, because, he says, "it was too inaccurate." In those days, tractors, chain saws, and trucks used leaded gasoline. Sometimes he felt faint from the gasoline fumes, but he paid no attention.

He bought a new farm, down in the valley, still holding on to 30 acres of the old. Because the buildings on the new farm were in sorry shape, he purchased them and the six acres for a bargain $3,000. "There was nothing but work here," Richard says. That could have been a statement in a poem by Robert Frost about farming the stony hillsides of New England. Richard worked, as he was accustomed, during all those hours that provided sufficient light.

He was occupied with a lot of physically exhausting bull work, even without considering the hard, outdoor, physical work of taking down and putting up barns. Richard's livelihood depended on his being strong and tough, able to endure whatever was thrown at him, a Marine as contractor. He was his own boss, without provision for time off, sick days, or medical insurance.

For ten years of dismantling and moving barns and reconstructing them as homes, his body responded admirably. Then, in 1968, "all at once," as he remembers it, his muscles began to fail him. He thought he was just tired. One day he couldn't keep down any food, and the next day he could not even manage a cup of tea. His hands and feet felt cold and numb. Worst, he was hyperactive, high-strung, and couldn't sleep. These were tough days for Richard—and tough on Donna.

He denied that he was sick and refused to go to a doctor, who might find out something really scary. Finally, after several days of complete disability, Donna drove Richard to the Berkshire Medical Center. Dr. Richard Lynch immediately confined him to the hospital and began a series of tests.

Among the questions he asked Richard were those relating to his exposure to lead, such as painting or automotive work. Richard truthfully answered that he had not had an unusual amount of exposure—not thinking about breathing the fumes from the tractor and the chain saw.

Diagnosis was difficult. Lynch considered hepatitis and multiple sclerosis, and finally, having ruled out alternatives, worked his way around to lead poisoning again. He ordered tests that showed extraordinarily high levels in the blood. The treatment consisted of calcium disodium edetate through the IV tube that was feeding Richard. The disodium edetate, with an excess of calcium, bound and sequestered the lead, which Richard passed through his urine. After three days of treatment the test indicated still a lot more to go.

Finally, after a week of treatment, Richard was able to sleep for five minutes. "How great that sleep was," he says. The next day he was able to sleep for five minutes again, and he began to feel that he was on his way. On the third day he was able to take two cat naps, and then the naps became longer. Gradually he calmed down. He could eat on his own, without the IV tube.

Yet he came out of the hospital in bad shape for anyone, but particularly critical given his occupation. He could not walk — in fact his ankles are still rubbery. "I've gone through a lot of pairs of shoes," he admits, wearing them down on the sides. It was a year before he could get back to work, while Henry finished up the job on which they had been working when he took sick. Richard was able to supervise.

Gramp, now blind, was also unable to work. He had always loved railroads so Richard, on crutches, drove his grandfather and his son Charley to

Steam Town, then in Bellows Falls, Vermont, where they all had a wonderful time riding the steam engine. After that success the three of them set out for Florida, to visit Disney World. That trip was a high point, especially for Gramp. Furthermore, the ticket takers in Orlando, confronted with a party consisting of a cripple, a blind man, and a five-year-old child, refused to let them pay admission.

Clayton Babcock died in 1973, at age 89. Richard dedicated the second of his self-published books to Gramp, through presenting a wonderful series of family photographs. A wiry Clayton Babcock stands in the middle of a bridge, in the winter, a pair of bib overalls over a plaid, wool shirt, clean-shaven, topped off with a warm wool cap. He looks, in short, just as he should.

Putting Together History

Meanwhile, frustrated by not being able to work, Richard began to haunt registries of deeds. Now he was seriously tracing back the ownership of barns—not an easy task, since outbuildings often were not mentioned in legal papers. Richard had a great advantage, however, as he sees it: he was untrained in what he was doing. So employees at the Northern Berkshire Registry of Deeds, lawyers, genealogists, and surveyors who were working there always seemed willing to help him. "If I'd gone to college," he says, "I don't know whether I would have been willing to show how little I knew." But when he knit his brows and looked puzzled, this semi-invalid drew assistance.

He remembers vividly the breakthrough when someone told him that if property passed to the immediate family, there wouldn't necessarily be any record. "That answered a bunch of questions I had," he says. He figured out that if he visited old cemeteries, where the names of original owners or leaseholders would be etched in stone, it might help bridge gaps in the registry records. His studies led him farther afield, to Pittsfield, the shire town of Berkshire County, and Albany, the capital of New York State, all the time photocopying the deeds. "It is beautiful to put together history," he says.

Something happened to Richard in the course of his research. "All that I was trying to do became greater," he says. He saw that his work with barns was a way of rediscovering history, or maybe even discovering it, from the perspective of barns. He began to read everything he could that related to colonial history, at least where barns might be mentioned. He visited the Sterling and Francine Clark Art Institute. There in books in the library he found seventeenth-century Dutch paintings that showed barns in the background, some with their doors open so he could see inside. In those books he found depictions of European barns, the prototypes of the ones he found in this country.

Looking back now he says the lead poisoning, terrible as it was to an active man, opened a new door. For the first time he began, regularly, to establish the provenance of the barns he moved. He discovered that history was not an abstract school subject, but tangible—found in the post holes more than in the textbooks. He began to understand that different characteristics of early barns related somehow to the nationality of the builders. He came to see that what had been written on that subject, although enthusiastic, was often inferior to hands-on experience or ownership research.

Richard was still hurting from his illness and still unable to do physical work. One day he was feeding the chickens. The coop was a terrible mess, he noticed. "Now how the hell am I going to clean that out?" he asked himself. He picked up a shovel and took a swipe at the manure, managing to toss it out the window. Then he took another. And another. Suddenly he realized, "I'm going to get this done!" He started to sweat, the result of his first exertion in three years. He shoveled that chicken manure, all of it, out the window.

And then, he says, "I thanked God."

As he writes in the acknowledgment section of his book *Barns in the Blood,* "I thank first my god who abides in me always."

VI

PUTTING ON
SOME ENGLISH

FORTUNATELY FOR RICHARD, interest in his barn conversion work did not die off during his convalescence. Word of mouth continued to bring clients to his door. Interviewed in 1974, he described how a barn home typically begins. When a buyer approaches him, he asks what features the buyer wants, how many rooms he needs and whether or not he wants a cathedral ceiling in the living room. "That gives me an idea of what size barn will be needed," he explained.

He consults his inventory, which includes standing barns farmers have offered as well as beams he may have at his shop. "I do not cut beams that are structurally important or in any way change the basic original construction," he said. "When the house is finished, it looks like a barn except for the fireplace and the rugs on the floor."

A barn restoration is not a cheap form of housing. The barn or barns must be taken down and the frame repaired. The building must be re-erected on a foundation and then, in effect, a second, stud-wall house must be built around it. A roof is required, and the interior must be finished. In 1996 dollars, a conversion runs $100 to $125 a square foot, more than a modest structure built from new material. That figure does not include the price of the barn, which could run anywhere from $1 to $20,000.

The clients for whom Richard has worked, however, want something more than utility. They want the shape and feel of the barn; they want to see the hewn beams and the complex yet sturdy joints. Just as many people today live on land that was once worked in agriculture, barn-home residents live in a structure that still speaks of the labor it contained. They live with an honorable history.

Some owners want the original siding put back on, with an exterior wall outside of that. If customers prefer to see the old barn siding, he wraps that with rosin paper, adds the stud wall, the insulation, the utilities, and the exterior sheathing.

Some want the interior to be lighter. After installing Sheetrock to the outside of the beams, Richard builds a stud wall outside of that, adds six inches or more in insulation and then wraps the structure with polyethylene or, now, Tyvek. Plumbing and wiring go in the new walls. Then he sheathes the exterior. This method is superior to cutting Sheetrock around the beams, which inevitably leads to air leaks, Richard points out.

For the Richard Lattizzori family, Richard moved a barn to Williamstown in 1975. His research showed him that original settler Amos White built the barn around 1750 in Lanesborough, Massachusetts, which was then called New Framingham. Like most of Richard's clients, the Lattizzoris wanted a cathedral ceiling in the living room, which requires more heat in the winter than does a room with a conventional ceiling.

Aside from extra heat requirements of cathedral ceilings, Richard's methods create barn houses comparable to conventional homes in terms of energy efficiency. Living in one of Richard's buildings is not like "growing up in a barn."

For his conversions, Richard had a truck adapted with a crane mounted on it. Although he uses a gin pole for his larger and more historically authentic projects, on more typical jobs the truck can lift the plates and rafters in place and save labor costs. He used the crane on the Lattizzori house.

In 1975 a slim, sturdy young man with blue eyes met me on my porch one sunny day. I said: "What can I do for you?"

He replied, "I'm David."

With a smile broad as could be, I said, "My God, I didn't even recognize you." I threw my arms around him, with his arms around me.

After some ten years of not seeing my son, it was a day I will never forget. We had many great days together getting acquainted again. The Lord must have looked after him all those nights I lay worried.

Not long after, another appeared—Hannah, my daughter. I didn't recognize her, either. She was beautiful; and with a boy friend, no less. Another great day! And soon after, Alice and Susan, two peas in a pod, arrived to do acrobatics for old Dad. They were beauties, too. Then came Charlene, a picture of me, as happy to see me as I was her. All spliced together, we had a new start.

Dave came to work like a tiger, so full of pride. He was strong. He made me proud. I had to slow him down without hurting his pride. He thought at first he had to just knock the barn down. It wasn't long, though, before he learned to save.

"I grew up in Pittsfield," David says. "Although I saw my father a few times, Mom generally discouraged it. They get along better now. I was always curious about my father. When I knocked on his door that time, he was headed out to a job. I had some carpentry work lined up, but when he asked me that day if I wanted to work for him, I took him up on it."

David worked for his father until 1983, when he and Clayton founded Babcock Brothers, a firm in Lee, Massachusetts, that works exclusively in barn restoration and conversion. Clayton, who trains hard as an amateur body-builder, later left the partnership but remains an employee.

The long shifts could still be stormy. "If people were goofing off," David remembers, "Dad would lay into them with colorful language. In part it was his time as a Marine drill sergeant; in part it was coming from a farm background, where 12- to 14-hour days were expected."

One man refused to leave the site when Richard fired him. Richard drove him off by throwing stones at him. The next day, however, the repentant fellow returned to work. David explains that, after starting at 7:30 or 8 a.m., around 4:40 or 5 in the afternoon some of Richard's employees were ready to leave the job. To his father, that was like saying they weren't devoted to barns.

"People say I'm a lot like my father," David says, "too much so to work for him forever. I like to do things my own way." David never intended to get into barns, but he inherited his father's fascination with how things were done in the old days. Now Richard sometimes helps his sons find barns.

Although Richard's younger children had gone along with him to jobs to help out, just as he remembered hammering nails with Gramp, David was the first son to return to be a real worker. The others followed over the years, so that eventually all of Richard's sons have had a go at barn raising. All except Charley, full brothers and half brothers, have worked for David, as well. Charley works with his father. He is quick and quiet, like second cousin Charley Haley.

The Eusden Barn Home

In 1976, Williams College Chaplain Reverend John Eusden and his wife Josie, then with the Counseling Center of the Berkshires, decided to build a home on a cul de sac the college opened off Bulkley Street in Williamstown. They had seen and admired a Babcock barn house, and so they called on the services of the barn restorer. Richard, who by this time had moved more than 36 barns, drove them about the area showing available, standing barns. The Eusdens chose a barn in Poestenkill, New York, which dated to 1750 and another, similar one, in Eagle Mills, New York, dated 1823. Both were English gunstock barns.

Richard, the Eusdens, and Wilson Ware, a designer from Sherman, Con-

Raising the first bent for the Eusdens' barnhouse in the snow in Williamstown, Massachusetts. (Josie Eusden photo)

necticut, worked together to create a home around the shape of a barn. They took the basic design from the older barn at Poestenkill and used many parts, including major beams and siding, from the Eagle Mills barn.

Richard worked on the project with his sons Clayton and Charley (and other sons for shorter stretches), Filmore Baker, Henry Suyden, and recent Williams College graduates Richard Deslauriers and William Flynt, who apprenticed with him. Deslauriers is now a contractor (Deslauriers Designs) in Franklin, Tennessee. Flynt is now architectural conservator at Historic Deerfield Inc., in Old Deerfield, Massachusetts.

The Eusdens held a barn raising after the bents had been completed. The late November 1976 day was far from ideal, however. Josie remembers serving cocoa and doughnuts in a snow storm to the neighbors and to Richard's crew, who raised the frame with a gin pole.

The Eusdens are grateful that when they had any trouble with their parts of the project, such as putting in the floor or building a stone wall, Richard volunteered to come over on Saturdays to show them how to do the job. The building code required that the floor be made of new material, rather than using the ancient flooring — a disappointment to the Eusdens. The house was built on a concrete foundation, to conform to the building code, but the stone from the Eagle Mills foundation was meticulously hauled to Williamstown to build the heat-reflecting wall for a sunken garden, on the south side, because the Eusdens had spent time in Kyoto, Japan, and wanted to reproduce a Zen garden.

The Eusdens say the costs ran higher than anticipated, but they love their house. The wooden beams remind them of the construction of Japanese temples — even the beam that unexpectedly ended up in the middle of the living room. They note that Richard walked about during construction with the plans jutting out of his back pocket, but they never observed him consulting them. "Richard told me I'd come to love that beam," John says. "And you know? I have." He explains that as a social gathering wears down, it is convenient to lean against that post.

Richard likes the use of space in the Eusdens' barn house and continues to bring prospective clients by to see it. Photographs of the Eusden home have appeared in several articles about Richard's work.

Widening Interest

Deslauriers liked his role as apprentice so much that he wrote the editors of *Yankee* magazine and suggested that they send someone to do a story about Richard Babcock and his barns. Lawrence F. Willard interviewed Richard at his farm on the Hancock hill for an article that was published a year and a half later, titled "Take Down a Barn, Build Up a House." The appearance of Richard's first widespread publicity, in February 1976, greatly broadened his reputation and brought him numerous jobs.

The piece struck the note that Richard was saving barns that would otherwise disintegrate or be knocked flat. "They are in the way, very often, of housing developments, shopping centers, and new highways," Willard wrote, "or they have become an eyesore too expensive to restore. To Richard W. Babcock, …however, old barns are part of our heritage, too valuable an example of early craftsmanship and honest toil to be allowed to pass from the scene."

In the story, Richard was quoted as saying that there was a good supply of barns because regulations had been pushing the "little fellow" out of dairy farming. "A lot of them couldn't afford dustproof ceilings, bathrooms in the barn, and that sort of thing, so they quit. When the cows went out of the old barns and there wasn't the body heat in there any more, the foundations began to cave in." Richard had tried to save what can be saved, before the timbers rot.

He said that he and his crew could take down an old barn in about two weeks but it took three weeks to a month just to restore the beams that were bad. Then the re-erection would go very quickly. He added that he believed that some old barns should be preserved as barns, but he was "very happy to be able to restore them to a long and useful life and at the same time preserve the craftsmanship of those early builders for all to see."

MSPCA Barns

The *Yankee* article led representatives of the Massachusetts Society for the Prevention of Cruelty to Animals to contact Richard. They visited him in Hancock in 1978, inspecting a barn he had erected to be his shop. "They told me they had seen others' work," he says, "but none could compare with mine."

He got the assignment to move two barns from Shrewsbury, Massachusetts, to Framingham to become the education center at Macomber Farm, MSPCA's working showplace. The barns belonged to Everett Harrington, who was descended from one of Shrewsbury's original settlers. This was Richard's largest project to date, and his first involving major dismembering, moving, and resurrecting barns to be barns again.

He could not have accepted the job, however, if Williamstown realtor Phil Alton, who had earlier recommended Richard to the Eusdens, hadn't intervened with the bank to get him a loan. The MSPCA could not pay the money up front necessary to get the project started, while Richard had to pay his crew for the weeks of work dismantling and moving the structure.

The earlier of the two barns, constructed of oak in the seventeenth century, was an

Richard's crew boarding in the English barn they moved to Framingham for the Massachusetts Society for the Prevention of Cruelty to Animals. (Jack A. Sobon photo)

example of a roots barn—one built in this country after the manner of the builders' native country. In this case, the builders were clearly English, as Richard could tell from the gunstock posts and beams dovetailed into the plates.

The larger MSPCA barn was much later than a roots barn. The doors were at the ends of the building. The smaller, older English barn had its doors in the sides, which was the custom. The English siding was always put on vertically, fastened to the sills, plates, and center girts.

Richard believes that the smaller English barn was originally built to store wheat somewhere in the valley of the Charles River. It was moved to Shrewsbury to be a dairy barn after the Revolutionary War. He was not able to research the records but could tell it had been moved. Richard's crew carried the foundation stones with the barn from the Shrewsbury site, laying them up in Framingham without mortar.

"We did everything," he says, "even the stone walls around the yards to the barns, the gates, and the landscaping. In other words, we moved the major

The English Barn

Gunstock posts, which those of English descent also used in houses, were created by inverting the tree, so that the wider part was at the top, leaving a bulge at the inside to provide space for two joints with different cross pieces. *Dovetailing,* a design associated more with furniture, increased the strength of mortise and tenon joints by cutting them in the form of one half of a bird's tail. The English barns were constructed with *trusses* or *braces* on each of the *cross beams* or *collar ties,* to strengthen the roof.

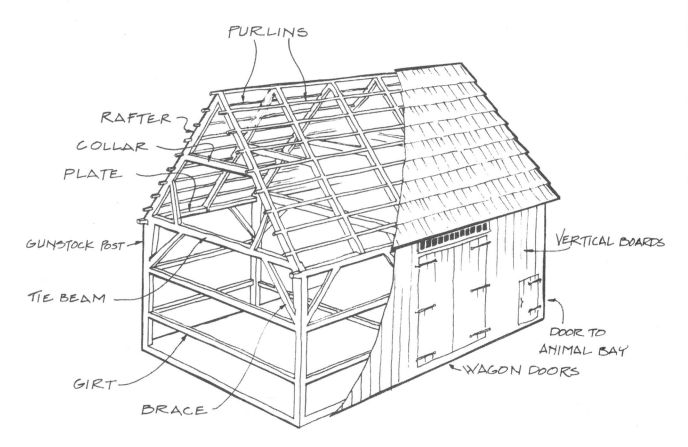

ingredients of the farm so that it could be understood in its new location."

Macomber Farm wanted barns of two different eras so that it could demonstrate the evolution of design and building techniques. "I installed the original cow stall and horse stall," Richard says, "after doing some detective work to figure them out." He was able to do that because he had grown up on a farm and because he had researched barns. The MSPCA people were impressed.

He worked with a large crew: three sons, Jack Sobon, Owen Grant, Bill and Bob Darling, Henry, and a man he hired in Shrewsbury. This project taught him how to work on a site more than a commuting distance from home. It taught him how to judge whether assistants were up to long-term assignments and how to work with a variety of building departments. It was covered with a major article in the *Boston Globe* and on Channel 4 Television, leading to other projects in eastern Massachusetts.

Some English barns Richard moved to become barns again and others he moved to become homes. In each instance he has found something unusual, something telling. In moving a gunstock post barn a short distance in Uxbridge, Massachusetts, for example, Richard found the date 1715 on one of the braces. Even the expert likes to have his estimates of age confirmed by hard information.

The Stiles Barn

In Southbury, Connecticut, Richard restored on site an English barn, built in 1673, originally with a thatched roof. The poles that held the thatch were set and pegged into oak rafters. The foundation under the north end of David Stiles's barn had settled 12 to 18 inches, pulling some of the joints apart. Dry rot had set in, as well. In 1985, Stiles's son, Ben, read an article in *Time* magazine about the Hancock restorer, under the headline "In New England: A Barn Is Reborn." The article, primarily about moving a barn from Eagle Bridge, New York, to Fairfield, Connecticut, quotes Richard.

"I'm an evangelist, truth to tell," Richard admitted. "Some men are called to save souls. I was called to save barns." The author had been impressed by Richard's judgment that the Eagle Bridge barn had once held sheep, based on the low position of a former hayrack. The reporter christened him "the Sherlock

Dave Stiles looks on as his English barn is raised on cribbing. Stone will replace the cribbing. *Opposite:* Henry Suyden (foreground) and Richard hew the same beam for the Stiles barn. Replacement parts were cut from descendants of the trees from which the barn was built 300 years before. (Georgia Sharon photos)

Holmes of barns," a title Richard enjoys.

After a call from Stiles, Richard visited the barn and determined that it was another roots barn. He was also pleased to be working for the first and so far only time with a descendant of the original owner of a barn that is still in use by the family.

Richard and his crew jacked up the north end and began to lay large stones underneath. Then they replaced deteriorated beams and rafters. Richard cut trees in Stiles's woods to replace the beams, hewing them out with his scoring ax, working with his son Charley, Henry Suyden, and Bill Hoffman, then a seventeen-year-old apprentice.

The next step was to replace collar ties that Stiles said his family had taken out in order to install a hay fork, a device that lifted hay from a wagon and carried

Richard had to replace the collar ties connecting the rafters near the roof in the Stiles barn in Southbury, Connecticut. These ties were frequently removed, weakening the barn, when a mechanical hay lift was installed. (Georgia Sharon photo)

it along a track in the roof to a loft. Hay forks, part of early mechanized farming, were frequently installed that way, which weakened barns.

Richard observed that the barn had been moved, because the original builder chiseled a "north" arrow on a beam that now faced west. Stiles said indeed that was the case. His family had owned a rock quarry. Cut stones had been used when they raised the foundation to keep animals underneath. The barn was rolled from its former flatland site, up onto the stonework. The big door, which once faced south, now faces east, while the animal entrance below faces south. All the timbers in the old section are hand-hewn, while those in a later addition show saw marks. Stiles's barn is now on the National Register of Historic Places.

Animal Rescue League

Another highly publicized move brought a barn to the Animal Rescue League of Boston's facility in Dedham in 1987. According to a National Geographic Society news feature on the barn, Richard referred to the project as "a fusion of history with modern methods for the treatment of animals." It was a barn to be a barn, the kind of restoration for which Richard has a particular fondness. He knows that barns have always been moved to be reused.

Enter Eric Sloane

Richard lauds Eric Sloane, author and illustrator of *An Age of Barns* and numerous other books on old buildings and old tools, for the interest he has created in old building techniques. He was looking at barns in the early 1950s, the last days when a significant number of roots barns were standing.

Sloane has taught the American public more about barns than anyone else. His books helped raise public awareness of their artistry and plight. "While I was learning from Gramp," Richard says, Sloane "was drawing shapes of barns and loving them" for themselves and as emblems of a way of life that was fading.

But, in Richard's words, "Sloane had no first-hand experience." That is, he drew conclusions about barns and building from what people told him. In some instances he had good information, sometimes not so good. He never cited his sources, so there is no way for anyone to check specific details. Richard says Sloane got much of his knowledge from the diary of an early Connecticut settler. Although the information on farming was full and good, only a small number of pages dealt directly with putting up a barn.

Richard Babcock and Eric Sloane finally met in 1979. The Davidson family had asked Richard to move a German barn for them from Route 22 in Petersburgh, New York, to Old Lyme, Connecticut, and they asked Sloane to draw it first in its original location. Sloane drove up and met Richard on the site. The two men walked around and talked a bit, and then Sloane said he wanted to get something to eat. Richard offered to take him to the Howard Johnson's in Williamstown, but asked him if he wouldn't like to see the inside of the barn first.

Sloane said he wasn't really interested. Richard was nonplussed, figuring that in order to know what a barn looked like, you had to know how it was built, as you might study human anatomy to before painting portraits. So he insisted. Sloane acquiesced — and perked up when Richard pointed out the joinery.

At the restaurant, waiting for their meal, Sloane began sketching on his place mat. He shoved it over to Richard. "Is that about right?" Richard was amazed at the likeness. Sloane took it back and began adding details, "most of which were accurate," Richard allows. He did help with some. Then the waitress came back, saw the place mat, and said, "Why, you must be Eric Sloane." Sloane said he was and gave her the place mat, to Richard's chagrin.

Sloane said he would rather see a barn fall in place than be salvaged. "I believe that's true," Richard says, "if we're talking about taking the siding and maybe some of the beams for interior decoration in someone's apartment. I told him I wanted to put barns back together, so they could live and be appreciated. It's true that being homes for people isn't quite the same as being barns again, but my idea was to move the history with them." When he moves a barn, he has done the historical research, which he passes on to the new owner.

Although a barn in ruins has its charm, Richard is hurt by the scene more than he admires it. And, he points out, barns in the way of a road or a housing development aren't left to molder in picturesque fashion. The bulldozer makes short work of them.

"I appreciate how historians feel about barns being moved," Richard says. "Barns would be better restored where they are, but that will not happen." Richard makes a distinction between Sloane and John Fitchen, author of *The New World Dutch Barn,* on one hand, and most other writers on the subject. While the first two include errors, they have "honest hearts," caring about their subject. Many of the other authors' books he dismisses as "mostly pretty pictures."

And the oak, German barn that Eric Sloane drew and Richard Babcock moved for the Davidson family stands today, an imposing structure looking out over the mouth of the Connecticut River.

VII

"MY FINEST HOUR"

IN 1966 CATHERINE FILENE SHOUSE donated 100 acres and $2 million to build the Filene Center, a soaring, 7,000-seat, open-air theater, at her beloved Wolf Trap Farm in Vienna, Virginia. Wolf Trap is the only national park for the performing arts. In 1980 Shouse, the daughter of an owner of Filene's Department Store in Boston and the granddaughter of the founder of the Boston Symphony, decided she wanted a smaller, enclosed performance center and reception center, to be built in the form of a barn, at the site. Wolf Trap had been, after all, a farm.

Conductor Sarah Caldwell introduced her to architect Mary Otis Stevens of the Design Guild in Boston. Stevens was interested in old barns and had read of the the work of Richard Babcock of Hancock. A student of early American vernacular architecture, she knew that ancient barns were related to medieval buildings and that barns in both the Old World and the New were similar to simple medieval churches. Hence she believed barns were proper places for public assembly and likely to be acoustically good.

Stevens persuaded Shouse to rebuild old barns rather than building a new one. Since most old Southern barns were destroyed during the Civil War, she asked Richard in 1980 if they could explore his inventory of Northeastern barns. The assignment turned into his biggest, most exciting and, even at under $60,000, his best-paying job up to that date.

In the summer of 1977, John Wolcott had written two articles for the *Altamont Enterprise,* essentially calling Richard a scalawag from Massachusetts who took New York State's heritage Dutch barns and sold them to wealthy inhabitants of New England. Wolcott is active in Albany area historic preservation. Richard, his skin thin to such criticism, was upset. He thought the comments unfair, and he was angry that they appeared in the paper without the author having contacted him directly about his misgivings.

Bill Flynt, working with Richard on the Eusden house that summer, suggested writing back to the newspaper, but Richard didn't feel comfortable doing that. So Flynt offered to write for him. In two letters he pointed out that the barns Richard moved were falling apart. Richard was doing a service for the farmer, Flynt wrote, helping him get rid of a tax liability. Furthermore, the barns were repaired for future generations to appreciate, even if not in the same location or serving their original purpose.

Richard had moved barns for New Yorkers as well as New Englanders, Flynt continued, and few of the barns were of Dutch origin, as those were generally too big for a house. Flynt's letters, signed by Richard, speculated that Wolcott, active in historic preservation in the Albany, New York, area, was criticizing Richard's work as a way of fund-raising for barn preservation. One of the letters notes that Richard was trying to make arrangements to meet with Wolcott in hopes they could coordinate their efforts. Later an accidental meeting took place, at the Albany Registry of Deeds, and it was warm and friendly.

The Kniskern Barn

Harold Kniskern of Blenheim, New York, on the east bank of the Schoharie River, saw the letters and called Richard. He said he believed Richard was doing farmers a service and invited him to inspect a barn built by one of his ancestors. Shortly thereafter, Richard came to take a look. And when Mary Otis Stevens called with the unusual request for a barn suitable for a 300-seat auditorium, Richard took her to visit the Kniskern barn.

He was able to recreate for Stevens the way the 40-foot by 50-foot barn, with a 30-foot peak, must have been laid out when Johann Peter Kniskern had it built early in the eighteenth century. The farm was in an area that was so fruitful that it came to be known as "the breadbasket of the Revolution"—that is, where the farmers grew the wheat that fed the Continental Army.

Upon entering the hay doors, the swing beam was to the left. This mammoth horizontal timber, a characteristic of German barns, measured 14 inches by 18 inches in cross section and spanned 38 feet. The swing beam made it unnecessary to have center posts in that section of the barn, while helping support the hayloft. The farmer could thus drive his horses and wagon inside, unhitch the team, and more easily swing the animals around to head them into their stalls.

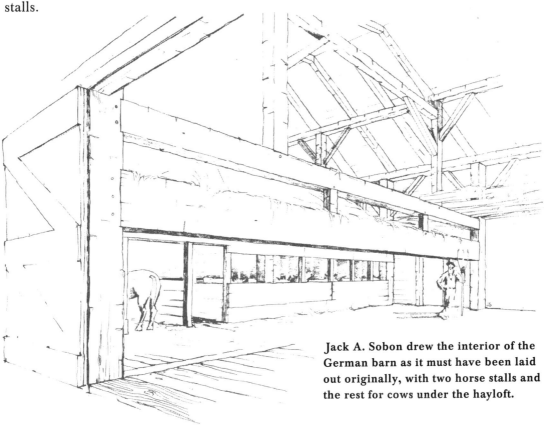

Jack A. Sobon drew the interior of the German barn as it must have been laid out originally, with two horse stalls and the rest for cows under the hayloft.

"The two horse stalls were to my immediate left," remembers Richard, "facing away from the threshing floor. The stanchions for seven cows, also to my left, were facing the threshing floor. Above the animals was loft space for hay. To my right, hay was stored from the floor to the roof. Above the threshing floor, small logs were laid side by side across the main crosstimbers, serving to store the wheat crop."

A detail illustrates the quality of the joinery and the hewing in the German barn from Blenheim, New York, that became part of a concert hall at Wolf Trap.

All of the beams were hand-hewn. The purlins and some wall studs were made from salvaged Dutch barn timbers, added later when a purlin system was built to support the weight showing in later years of the snow load on the rafters. The braces were sawed with an early up-and-down saw. The craftsmanship was excellent, certainly done by a master carpenter, Richard says.

The Kniskern Family

Johann Peter Kniskern and his wife settled in the valley in 1712, having come to this country with Palatine Germans, sailing from Rotterdam to London and then the New World. Their homeland had been devastated by wars; they had been heavily taxed and persecuted for their religion. The winter of 1708–09 in central Europe was so bitterly cold that wood would not burn in the open air; fruit trees and grapevines were killed.

Thus the migration. So many people from the Palatine state—along the middle of the Rhine River—were among the German emigrants arriving in London between May and December 1709 that all were called Palatines, even though some came from neighboring countries. The British government tried to help out with a proposal to start a tar works, based on pitch pines, for the British Navy in the New World.

When the emigrants actually sailed up the Hudson, however, they discovered only a limited number of pitch pines. The project was abandoned. Not only were the Palatines sick with diseases of the voyage, including typhus from which many died, but the British then told them they would have to shift for themselves since the deal was off.

Some, like Kniskern, dealt directly with the Indians for land. He was one of fifty families to settle along the Schoharie River in 1712. The Indians helped them through the first winter. Other Palatines tried to find work with the Dutch in the Hudson Valley. The arrival of this group in the area is the reason why German barns appear where the Dutch were thought to have been the first European farmers.

Johann Peter Kniskern was a "listmaster" and captain of twenty-four men from Hunterstown who joined an expedition against the French in Canada. He is buried near the barn site and the old family house. At least one branch in each

The German Barn

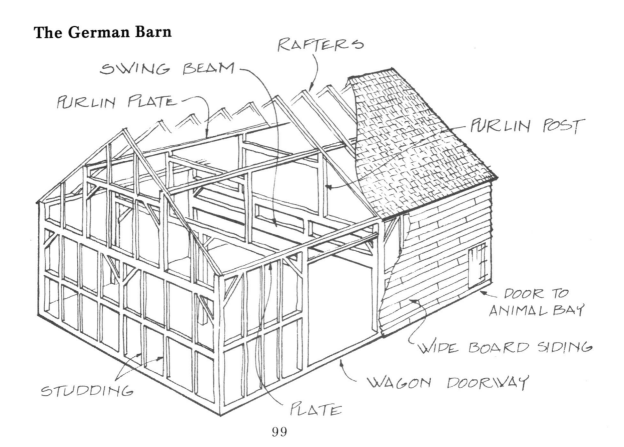

generation since has stayed on the land for which their ancestor fought. Harold Kniskern and his wife, Gertrude, were the last to make a living off the land.

The Valentine-Bump Barn

The other barn the architect chose was a first-generation Scottish barn, built in the 1790s along what is now Route 22 in the town of Jackson, New York, at the southern end of Lake Lauderdale. The three-bay barn, 30 feet by 40 feet, was originally constructed to be a flatland barn.

Upon entering the main floor of this barn from the highway (west side), I saw that the cows and horses were originally stabled to my right and the haymow had been to my left. Being a medium-sized barn, there was room for five cows and two horses. All the main timbers were hand-hewn, while the smaller side girts and knee braces were of up-and-down-sawed oak. The wall boarding was vertical and the upper ends were inserted into a groove in the underside of the plates and the end wall crosstimbers. This ingenious feature saved on nails, a precious commodity in the New World.

Above the center bay, or threshing floor, there had been poles stretching across, indicated by notches in the cross beams, for the drying and storing of wheat. When the slate roof was installed, extra rafters, hewn only on one side, were added to carry the added roof load. The roof, like its thatched predecessors in Europe, was steep, the pitch being 10 in 12. The purlins and plates were of pine and ran the full, 40-foot length of the barn.

The barn's original owner, Joseph Valentine (probably a descendant of the Lancashire, England, Valentines), was born in North Hempstead, Long Island, in 1750. He fought in the American Revolution under the command of Captain Swartwout. Because his Long Island relations were loyalists, he changed the spelling of his name to Volentine.

In 1790, Volentine traveled by ox team and wagon to Jackson, New York, where he settled with his wife and twelve children. He stuck the ox goads used on the trip into the ground upon arrival. They subsequently grew into great poplar trees.

The property was adjacent to Long Pond, now called Lake Lauderdale. In 1746, 900 French and Indian troops camped there en route to Canada with their captives from a battle at Fort Massachusetts, in what is now North Adams. In 1777, Colonel Baum and his German mercenary soldiers stopped on their way to an arms store in Bennington, Vermont. Instead they were met and defeated at the Battle of Bennington, actually fought in North Hoosick (Walloomsac), New York.

Although the Valentines were English, the Jackson area was settled mainly

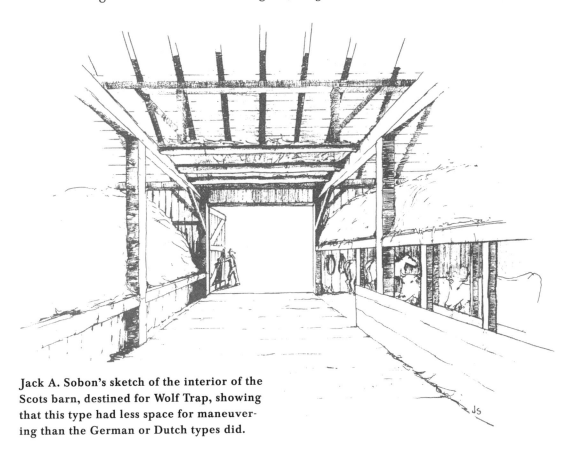

Jack A. Sobon's sketch of the interior of the Scots barn, destined for Wolf Trap, showing that this type had less space for maneuvering than the German or Dutch types did.

by Scots. Richard's studies indicate that most of the barns in the area were of Scots design and at least two seem to have been built by the same builder as the Valentine barn, based on details and craftsmanship. One of the neighboring barns was built by three McGeoch brothers in 1700. All of them were carpenters, as were some of their descendants. Perhaps a McGeoch or several built the Valentine barn.

The farm passed to the oldest son, Daniel, and then to his daughter, Rebecca, who married Horace Bump. Their son, Charles, was the last to operate it as a farm. The farm passed through three other pairs of hands before Richard acquired the barn. Both Charlie and his sister, Mary Elizabeth Fousnaugh, are happy to see their ancestors' barn at Wolf Trap.

Charlie Bump, who was living in Fort Edward, remembered an earlier time when the family barn was moved to a hillside by the highway to take

The Scottish Barn

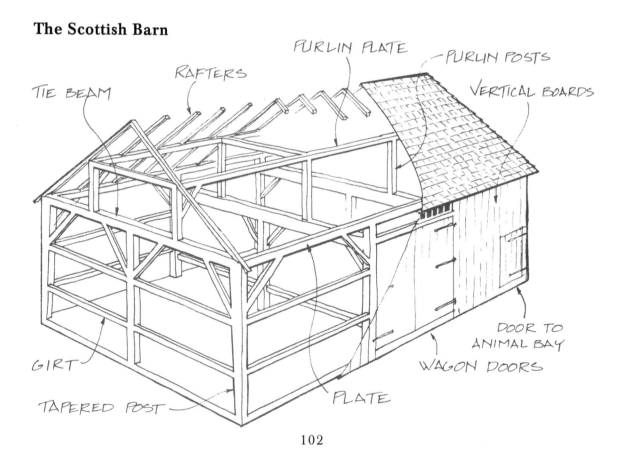

The Scots barn as it stood in Jackson, New York, prior to being dismantled and trucked to Virginia. (Jack A. Sobon photo)

advantage of the slope, so that animals could be housed in the lower level and hay stored above them. As Richard points out, builders had to trust their work with stone enough to lay a seven-foot foundation, without mortar—high enough to keep the cattle inside.

When builders moved the barn a short distance, he says, they stripped it down, stiffened the frame with planks, and pulled it over rollers with a team of horses. A drag horse would haul the rollers from the back around to the front. Once the barn was situated in its new site, a lean-to was added and the roof redone in slate from nearby Granville.

"People in the town of Jackson will miss the barn," Richard told a reporter, "but it isn't really gone because it will be kept alive at Wolf Trap, not like so

103

David Babcock adjusts
the rigging on the gin
pole while disassembling
the barn at Jackson
headed for Wolf Trap.
(Jack A. Sobon photo)

The partially disassembled Scots barn, with the gin pole raised against the sky.

many others that are gone, never to be seen again."

Richard took down the Kniskern barn in unexpectedly beautiful, temperate winter weather, and the Valentine barn in Jackson in early spring. While he was working by the road in Jackson, an elderly gentleman came along and fell into conversation, as often happens during barn work. The man said he knew that barn quite well. One story he told went back many years to when he was a child. He and his friend were walking nearby when they spied a large tub of ice cream, perhaps five gallons, fall off a truck. They grabbed the container and took it into that very barn, where they stayed eating strawberry ice cream with their hands all day.

Richard asked the man where he could cut a gin pole. The man invited him onto his own adjacent property, where Richard cut a hickory he used to take down that barn. He carried the tree to Vienna, Virginia, where he used it to raise the barns at Wolf Trap.

Transporting Barns

When Richard had completed dismantling both barns, he hauled the parts to his workshop in Hancock. There he made repairs during the next two months.

A caravan of trucks and cars drove south in early May. Trucker Ralph Dodge of Florida, Massachusetts, carried the timbers in his Freight Liner. Five of Richard's sons went along—David, Charley, George, Clayton, and Ronald. Ron was too young to work on the project but enjoyed the trip, especially talk-

ing on Dodge's CB radio. David stayed in Hancock to work on barn pieces in the shop. Other workers who went south

Richard anchors the raising team as the first bent for the German barn rises at Wolf Trap. (Jack A. Sobon photo)

were Henry Suyden, Jack Sobon, Paul Martin, Donald Zell of North Adams, and James Hoffman of Stephentown, New York, a descendant of the Palatine migration that created the German barn.

Reconstructing

Long Construction Company, hired by the National Park Service, had already built the foundations. Richard put down the sills and flooring and then began the frame. He worked directly for Catherine Shouse, part of her gift to the endeavor. One day, while they were using the gin pole to hoist the last bent, they heard an ominous cracking sound beneath the floor on which the gin, under the stress of lifting, was resting. A very public disaster seemed imminent. Richard would not hear of stopping, though, with a bent partway up. Fortunately, the floor held until they could get some shoring underneath. Then Richard took that same hickory gin pole, cut next door to the Valentine barn site, and hewed it to replace the broken beam.

The drill for the crew was to drive home late Friday and leave for Virginia at about 4 a.m. Monday. One Monday when they arrived as usual at 10 a.m., Shouse and members of the Wolf Trap administration were standing staring at the partially erected barn frames. Richard was worried that something was wrong. Shouse explained that the wind had blown so hard the day before that it had tumbled a 30-foot-high masonry wall near the site. "We were afraid that something might have happened to the barns, Mr. Babcock," she said. Richard was proud that his barn frames withstood a blast that could flatten masonry.

Catherine Shouse honored us with her presence on several occasions during the course of the raisings. She could also view the work in progress from her home there in Wolf Trap, which sits on a little hill above the site.

Gramp, how pleased he would be. Through us the knowledge of how to erect the barn in the old way had survived and was put to use—and, as well, the stories he told me of how the old timers taught him.

I sensed that this was a well-thought-out plan of Catherine Shouse's to provide a special place to join the past and present, and create something unique. It would be the dream of many aspiring performers to present their talents to the world here. For others, there would be the double pleasure of enjoying these performers and at the same time experiencing the "roots of America" in the barns themselves.

The raising of the barns was pure joy. It was a coupling of the confidence of our specialized experience with preserving histori-

cal structures, allowing the greatest pleasure and the grandest barn raising ever
in my life. The barns at Wolf Trap Foun-

As soon as both the German (left) and
Scots barns are raised, a fir tree is fastened to the rafters to celebrate—and
to honor wood. (Jack A. Sobon photo)

dation for the Performing Arts represent my finest hour, and a grand
and glorious period in the lives of all who were connected with the
project, a time that will live long and proud in our memories.

By the end of May, when all the frames and rafters were in place, Jack
Sobon nailed a small evergreen tree to the peak of the Jackson barn, a tradition
that recognizes a builder's gratitude to the wood with which he works. James
Hoffman tuned his guitar and the crew sang a few songs. Then they nailed the
old roof boards and siding back on.

The Queen Post Truss

As Richard's part of the project wound down and Long Construction took over to build the exterior shell and the interior, Richard became increasingly concerned about the architect's and engineer's plans to compensate for three posts that were removed in the Kniskern barn to give an unobstructed view of the stage. In response to his inquiries, three times the architect and engineer told him that a metal or cable truss would be installed before any weight was added to the roof. So Richard left.

He was happy when he got home, but the concern ate at him, too. Two weeks after returning to Hancock, in mid-June, Richard asked James Hoffman to jump into the car with him and head to Virginia. He asked Hoffman because he knew the young man would do it. When they arrived, men were on the roof of the German barn. There was no sign of the truss. The main cross beams were sagging.

Richard yelled up to the men to get off. "Can't you feel the roof bouncing when you walk on it?" The Long supervisor came around the corner. He recognized Richard and asked what was going on. Richard demanded where the trusses were. The supervisor said there were no trusses in the plans. He asked Richard what they should do. In retrospect, Richard is grateful to that man, who could easily have gotten just as mad at Richard as Richard was.

After measuring the distance to the floor and adding six inches because of the sag, Richard led the crew into the nearby woods to cut trees to shore up the cross beams. The men carried the three long logs on their shoulders back to the barn, then stood them up, as they would raise a flag pole. The supervisor told Richard Catherine Shouse was on the telephone.

Richard's ears stood up, just as they had for his grandfather. He figured he had gotten himself in for it as he picked up the telephone in the supervisor's field office. Rather than bawling him out for interfering, however, she invited him to her downtown Washington home to talk about it.

The butler showed him into her living room, serving tea and cookies while she asked him to explain the situation. He did. She asked him if he had brought up the matter before he left. He said he had. She asked him what the architects and engineers had said. He replied, "They told me my job was complete. They would take care of it."

Then she asked why he came back down.

He said, "Well, I have never lost a barn yet, and I darn sure don't want to start with this one."

She asked him to design a truss, with an estimate for its installation. Richard agreed. He also gave her prints of photographs he had taken during the raising. She later had them blown up and mounted on the walls of the reception area.

"When Catherine asked me to design a truss that would work," he says, "I knew it could be done, because it already had been done." One reward that comes from working with old buildings is the sense that the problems have already been met—and solved. The knowledge is already there. "Every barn I took down, or Gramp did, it was right there."

Richard returned to Hancock to sit down with Jack Sobon, who had been trained as an architect. Richard suggested a king post truss. Jack said that would be a lot of work and more expensive than was needed. He sketched a queen post truss made of old wooden beams. Richard says, "He was right and I knew it," so he sent the drawings to Shouse, saying he hoped they could do it for $4,000; if they ran into trouble, it might take $6,000. She approved the proposal and Richard returned with his right-hand man, Henry, to do the job.

When the queen post truss was in place, they knocked out the temporary shoring. No sagging. He billed her for $4,000.

Thus in the booklet that accompanied the Wolf Trap Foundation for the Performing Arts commemorative stamp issue in 1983, Richard is credited as "designer." Richard characterizes that notation as kind, Catherine Shouse's way of publicly thanking him. Richard does not think of himself as a designer, however. "Jack designed the truss," he says, "but in a way, we both simply reached into the past to come up with the best solution."

For more than ten years, Wolf Trap Farm, in the foothills of the Virginia Blue Ridge Mountains, has been a national summer center for the performing arts. Thanks to the recent addition of the

"Barns"—two buildings as old as they are new—Wolf Trap now functions year round.

The Barns, which date from before the Revolutionary War, originally stood in New York State. Both buildings were dismantled by Richard Babcock of Hancock, Massachusetts, and reconstructed in Vienna, Virginia.

Though he did the work in 1980, Babcock used nothing but 18th-century techniques to get it done—block and tackles, gin poles and pulleys, and manpower. He reconstructed the barns from the inside out, carefully designing modern additions to complement the hand-cut timbers and weathered boarding of the original buildings.

In the great tradition of barn raising, the newest addition to Wolf Trap Farm functions as an informal, congenial site for dance, music, film, opera, theater, conferences, and conversation. The Barns are a promise to the community of creative abundance and a symbol of Wolf Trap's legacy.

—*United States Postal Service, 1983*

Shouse thanked Richard more personally in a letter dated November 4, 1983. "Your Barns have been enjoyed day and evening," she wrote, "by many people attending the stage presentation, literally thousands of our Head Start activity groups, and by private groups who, in increasing numbers, find the Barns a perfect setting for weddings and other celebrations.

"You are in our hearts and minds as we explain the Barns and their construction to many interested people from our country and abroad. One of the framed collection [of the photographs Richard took] of the building's progress will hang in the Barns with your letter, and one in my library at Plantation House"—her Vienna home.

Richard wrote back that he was self-publishing a book about barns, which he wanted to dedicate to her. She wrote back July 15, 1988: "It is indeed flattering that you wish to dedicate your new book to me and I am happy that you want to do this. I will look forward to its publication. It will make a great contribution to Americana and I am anxious to see it."

Openings and Closings

At the official opening of the Barns at Wolf Trap, January 11, 1982, the Primavera String Quartet premiered a comedy show, "The Columbine String Quartet Tonight." At an unofficial test run, the No Elephant Circus had performed to the delight of a young audience a few days before. And before that, pianist Earl Wild and the London Savoyards gave private performances. The performers and the audience pronounced the acoustics to be excellent. Song, circus, and quartet were performed on a stage built beside Kniskern barn, in which the audience sat. The barn was attached to the Valentine barn by an understated L-shaped main entrance that gives access to both. Thus, for the first time, Wolf Trap was open year-round.

The Primavera played a more traditional program—Haydn, Stavinsky, and Debussy—on January 14. A Kurosawa film festival, mezzo-soprano Evelyn Lear, the Wolf Trap Chamber Group, and Trapazoid, an Appalachian string band, followed.

After the winter season, the barns serve as support for the outdoor performances. Other facilities they house include a courtyard, an orchestra pit, two principal dressing rooms, two chorus dressing rooms, a rehearsal space beneath the Scottish barn, and a box office.

Benjamin Forgey wrote in the January 9, 1982, edition of the *Washington Post* that "the Kniskern barn, built in the first quarter of the eighteenth century, makes an excellent, warm, intimate place for certain kinds of performances.

"The space feels right. It suits its new purpose like an old glove. The aged textures of those hand-hewn, hardwood beams and studs have something to do with this quality, but it's also due to the space itself, almost as if Johann Peter Kniskern had something more than hay and horses [and wheat—Ed.] in mind when he built the barn in the Schoharie River Valley near Albany more than two and a half centuries ago."

In the summer of 1982 a disastrous fire gutted the main auditorium at the Filene Center. That October Richard and his five sons planned to dismantle a Dutch barn at his Hancock museum, cart it to Vienna on a flatbed truck, and re-erect it as a fund-raiser. He called it "a gesture of gratitude to Catherine Filene Shouse." The proposed gesture, although appreciated, was unneeded. The theater was rebuilt with the aid of an $8.5 million federal loan.

Harold Kniskern had been proud when Mary Otis Stevens and Catherine Shouse chose his family barn to become part of the performing art center. In 1982, however, he read or misread a story in the *Schoharie Review* that said the Wolf Trap barns had burned. Richard arrived to visit Harold, who had been ill. Richard told him wonderful stories about putting up the barns and how good they looked. Harold's expression didn't change. Finally he said: "Too bad they burned," believing that Richard was stringing him along because he had been ill.

"Good God, they didn't burn," Richard said, explaining that the Filene Center auditorium had been destroyed but not the barns. Harold Kniskern was a relieved man.

Catherine Shouse died on November 16, 1994, at 98 years. In 1921, when she married Alvin E. Dodd, her father gave her $250,000. It grew to $1 million but dwindled to $9,000 in the stock market crash of 1929. She used $5,200 of what was left to buy Wolf Trap Farm, then 52 acres with a dilapidated, 300-year-old farmhouse. She augmented an $800-per-month allowance her father gave her by selling eggs and milk from the farm to her neighbors in Georgetown. In 1932 she married Jouett Shouse, newspaperman, lawyer, and Congressman who served as chairman of the Democratic National Committee and assistant secretary of the Treasury. He died in 1967. She continually purchased additional acreage for the farm, as it became available.

She rebuilt her fortune more with investments than eggs and milk; nevertheless, like Richard, she was a farmer. Like Richard's grandparents, she has passed on "to that world where the barns are always full of grain and hay, and the cows are full of rich milk," as Richard had said of his grandparents. She had the insight to see—and make a public statement of—the productive and inspirational relationship between agriculture and the arts.

The Barns at Wolf Trap

Here timbers tall align the wall
　　While others still by plate and sill
Where rafters mate in seats of plate
　　From where they rise to
　　Waiting skies
With beams and braces all locked together
　　We now have the hope they'll last forever.

In harmony men labored long hours in the sun
　　Hewing timber we see here all done
Like great windows out through the sash
　　We all catch a glimpse of yesterdays past

Timbers once hewed so far in the past,
　　Still show the mark of the timberman's craft
Raised in splendor all hoisted with pride
Reliving this history, we all took a ride.
Remember all these men of yesterdays past
　　When you see all the timbers
　　That weren't made too fast.

Ropes and pulleys pulled hard by the side
Raised up each section with honor and pride
Then raised to the rooftop, as it should be,
　　The symbol, evergreen tree.
The message was clear, our forests are dear.

Inside we see, the old and the new
A fusion of history and modern art, too.
Here something special was added in, too.
With timbers all crossed and braced from aloft
Where purlin post hide and tend to divide
　　Weight from above
　　Locked now in love
　　"A queen post truss"
　　History and hope
　　All put up with rope
　　Preserved here to shine
　　By a woman so fine.

—R.W.B.

VIII

IN DUTCH

WHEN LAWRENCE WILLARD WAS interviewing Richard for the first *Yankee* magazine article in 1975, the master builder was still suffering from the effects of lead poisoning. At the end of the interview, talking primarily about barn conversions, Willard asked him what he would like to do if he could do anything. Without consciously having thought of it before, Richard blurted out: "I'd like to build a barn museum."

In 1977 he began, moving a Dutch-style barn, built about 1750, from Guilderland, New York. The original owner was Mathias Groat. Richard continues to use the barn as his workshop.

While it was a good example of later Dutch barn architecture, it was not a museum barn, Richard felt. What was missing was the history. So he spent many weekends looking in areas where he thought a jewel might be.

"Finally I saw it," he says. "I was probably a mile away from it when I saw it for the first time. That's how well I knew what I was looking for."

The Great Plantation Barn of Saratoga

He was traveling north on the east side of the Hudson River when he came to a small rise. What he saw first, across the river, was the roof, steep and large. Next he noticed the square shape and low side walls, characteristic of Dutch barns. It took a while to find it; fortunately it was on a rise, or he might have had even more difficulty figuring out how to get there. Later he learned it was built on higher land as a precaution. A flood along the Hudson in 1646 had wiped out many barns and houses. As he turned up the drive to the farm, he was absolutely convinced it was the barn he wanted.

When knocked on the farmhouse door, he met Leo Thelavarge and his wife, Rose, who were pleased to show him the barn. The sills were gone and the posts had sunk into the ground. But, as is often true of the Dutch barns, the heart was still intact: the huge H's of anchor beams and supporting posts. That heart is the core of the barn. When the posts and walls have rotted, the core stands firm.

During his inspection Richard was overwhelmed by the barn. It was what

Richard's drawing of the barn he christened the Great Plantation Barn of Saratoga, in the condition in which he found it.

This detail of the massive anchor beam shows the loving work that went into the reconstruction of the barn at the Babcock Barn Museum.

he wanted and more. The owners, who had despaired of ever repairing it, were delighted with the prospect of having it moved. They wanted to build a small pole barn to meet their present needs. They were pleased that it would go to a museum.

Richard had no money to invest in the project, so he had to work "as efficiently as possible." He believed if he could just get it down and over to the workshop, he could make the repairs while doing his regular work. So he dismantled it in odd hours.

He carted stone from the old foundation and laid it up in Hancock. Somehow he found the time, before winter, working with his sons and Henry, to raise the main anchor beams on sills newly made of old timbers, splicing new wood onto the old posts. They put the old roof boards on again and rolled out roofing paper—all that he could afford. "I never did get on the wood shingles I wanted," he says. Later he took off time again from his paying work to close in the sides; still later he and the crew filled in the stone under the sills. The research, too, was arduous.

That was how, in 1978, he came to move the centerpiece of The Babcock Barn Museum, a 1685 Dutch barn from Schuylerville, New York, with timbers up to 50 feet in length cut from a single tree. The barn was originally owned by Jan Jansen Bleeker, who arrived in this country in 1658, at the age of 17, and later became a civic leader and mayor of Albany. Richard called this structure the Great Plantation Barn of Saratoga, as it was built on Saratoga Patent land. The barn was made entirely of wood, including

Richard believes the Saratoga barn, before it burned, was the first totally restored plantation barn in this century. ➤

118

the hinges on the huge doors.

He officially opened his museum, focused on this building, in the summer of 1983. Richard brought it to museum quality by re-creating the hinges and re-constructing the hayrack, the structure the Dutch invented for keeping a horse from wasting hay. The joinery was done meticulously, mortises and tenons drawn together by wooden wedges.

As well as two old wagons, Richard reconstructed a great apple grinder and cider press from a barn in Petersburgh, New York, for display inside. The machinery on the grinder was turned by either horses or slaves.

The Saratoga barn and its contents were what burned in the summer of 1994. The barn museum has not recovered.

Northern Slavery

His researches into the Saratoga barn took him in an unexpected direction, for in experiencing Dutch barns he came to realize that their size, the farms they serviced must have required slave labor. Although historians nowadays acknowledge that agriculture on the Dutch patents along the Hudson was supported by slave labor, to many people the extent and importance of slavery in what is now New York and New England come as a surprise. Richard says he never learned anything in school about slavery in the North; only that African Americans came north to escape slavery.

Richard remembers giving a talk on the Saratoga barn at the Schuylerville Historical Society. When he mentioned slavery, several people in the audience asked if he didn't mean "indentured servants"—immigrants who had to work off their passage. When he said he meant black African slaves, they found it hard to believe. Richard had the documentation, though: slaves mentioned in wills and other legal documents. Indeed, General Schuyler, who fought at the Battle of Saratoga in the Revolutionary War and for whom the town was named, freed his slaves on his deathbed.

Upstate New York and New England pride themselves on being stops on the Underground Railway that helped runaway slaves escape. Basements built to incarcerate slaves are something else. The year after Richard gave his controversial talk, the owner of the farm next to the Saratoga site broke into an enclosed chamber in the foundation of his house. There he found chains and whips used on the slaves who were forced to live without light or fresh air. These slaves produced the wheat that fed General Washington's army.

Richard has located numerous unmarked graves in the vicinity of barns. While many local people have regarded these as "Indian graves," local tribes did not bury their dead until the coming of Christianity. The slaves, so conspicuous in the wills, had to be buried somewhere. For barns to have been built next to Indian burial sites would have been a coincidence; for slaves to have been buried next to barns would have been a convenience. And what about those unmarked depressions in the cemeteries given over to the original founders of New York and New England communities?

Most of the slaves were buried in slave cemeteries. Because slave graves were more casually marked, if they were marked at all, they were probably built

The Unknown Slave

A wonder,
    You might see
    The marker white birch tree.
Upon a hill, and in a mound,
    There about six foot down,
    Still over-looking Hancock town,
    Lie the bleak remains
    Of one who came in chains.
His story should be told
    As the flower must unfold.
    Others like him have been swept away,
    With no mention of their clay.
The time was early with the land,
    And all he did, he did by hand.
    His days were spent full and complete
    Toiling at his master's seat
    Until at last his final sleep.

His only marker this nature's mound
    And lonely birch tree above the ground.
    Only a small depression in the soil
    Gives him rest from all his toil.
Perhaps some offspring yet unknown
    May one day find his ancestral bone.
    If by him his seeds were sown,
    Whereafter flown.
Is there still hope of coming home?
    These graves are seldom known,
    Deprived as they are of marking stone.
What chance is left this site be known
    For others connected to his bone?
    These bones so dry,
    No wonder why,
    Await his story told.
                —R.W.B.

121

over when the farms were moved out of the valleys in the face of development of communities. For easy digging, many slaves were buried in gravelly soil, Richard believes. When the towns wanted that gravel to build roads, they simply tossed the bones away. Where stones marked slave graves, often they were recycled to build roads and foundations. The topic of slave remains in the North has not been studied.

Jan Jansen Bleecker was one of seven partners who purchased the Saratoga Patent. Richard says from his studies, he knows what Bleecker did next. History records, whether we like it or not, that a slave owner's first order of business after surveying his property was to dig a hole in the ground eight feet deep and to lay stone around it to support the house. This sealed basement doubled as the slave chamber, its only access through a trap door into the first floor, so that the slaves would not run away at night. The slaves built it, as they helped with construction of the barn.

Philipsburg Manor

Another great Babcock Dutch barn still stands, open to the public, in Tarrytown, New York. In January and February of 1982, in weather as cold as any on record, Henry, Richard, and his sons David and Clayton found themselves disassembling an old plantation barn, a seventeenth-century Dutch barn in Guilderland, New York, built for the Oxburger family. The workers had to brave 30-degree-below-zero weather because the barn was wanted that spring to replace one that had burned at Philipsburg Manor estate, operated by Sleepy Hollow Restorations, now known as Historic Hudson, Inc.

"Today builders fight Mother Nature," Richard says. "The original Dutch barns, with their dry stone foundation and all-wood construction, were built to exist in harmony with nature."

Frederick Philipse originally built seventeenth-century Philipsburg Manor, which at the time included thousands of acres along the Hudson and Pocantico rivers. When the estate was originally opened to the public, in 1966, it had a Dutch-built barn dating from the 1720–1750 period, moved to the site from Old Hurley, New York. But in August 1981 an arsonist torched the barn, home to grain storage and thirteen farm animals bred back to resemble those used in the eighteenth century.

Sleepy Hollow Restorations regarded a barn as central to its effort to interpret eighteenth-century life, because the Upper Mills, now restored, had made Philipsburg Manor an important trading and agricultural center. Tenant farmers came to pay their rent in wheat, which was ground and sent to New York City and abroad as flour. Someone mentioned a "barn man" to John W. Harbour, Jr., then executive director of Sleepy Hollow. Harbour conceived of raising a new, old barn in April at the site in the old way—with workers wearing

The Dutch Barn

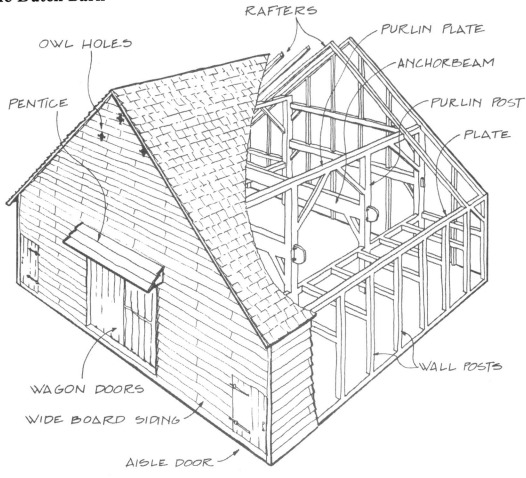

Starting dismantling the Dutch barn at Guilderland, New York, in sub-zero weather.

eighteenth-century costumes—as part of the interpretation of the property.

The barn Richard chose was on the old Ogsbury farm and was slated for demolition to make way for a housing development. Richard traced the deeds back to David and Elizabeth Sheldon Ogsbury. He could go no further. Then he discovered that the name had been changed from Oxburger at the time of the Revolution. As the barn came apart, Richard sleuthed out that it had been moved from still another site, probably the Hudson near Albany.

In the extreme conditions at Guilderland "we used no fires at the site, no heaters, we just took the cold," Richard says. "If you build a fire [on a job site], people hang around it rather than working. Everyone adjusted to the sub-zero weather the right way, with hard work. Each man covered his frame first with longjohns, then more layers of whatever else suited him—enough so he could, after warming up, take off a layer or two."

124

On January 15, 1982, David Babcock made the hard climb to the peak of that roof, just to peel up a metal cap, which

The prow of the Dutch barn seems to be sailing across the snowy fields of Guilderland in the bitter cold. (David Babcock photo)

marked the beginning of the work. Suffering a wind that wouldn't quit at 30 below, he peeled that metal strip, which had been nailed hard to the peak. We saw that his face was red as a beet, but trying to get him down to thaw out was a futile effort. Soon the cap slid down the roof.

Hard winds lowered the windchill many days to 50 degrees below zero as we carefully peeled down the frozen mass at Guilderland, south and west of the Normanskil River. The sun, even between the hours of 11 a.m. and 1 p.m., did little to warm the freezing limbs as Henry, David, Clayton, Vern Tower, Todd Framback, and I worked to save this barn.

The rule of disassembly is simply the reverse of assembly, whereby the last items installed become the first to be removed. The last item installed was the roof. In this case, however, there was more than a seventeenth-century Dutch barn. The old wooden roof had been covered by slate, the interior had additional floors, new siding and many windows had been added.

My aim was to disassemble all twentieth-century additions, leaving only the original materials so that we might see what the barn looked like when it was originally put up—the distinctive H of the anchor beams at the heart of the Dutch barns. I wanted to photograph the disassembly of the original barn, so that we could have an account of how a seventeenth-century Dutch barn could be moved. So the roof was the first to go.

The rafters appeared as we stripped the roof of slate, wood shingles beneath, and boards beneath them. The rafters were original, first handcrafted with scoring axes and then hewing axes. They were joined together with wooden pins. What a day when the sun for the first time in hundreds of years struck the old ax marks.

Then we removed the twentieth-century siding and windows, exposing the handwrought posts and beams. The spaces between the posts were insulated by being stuffed with hay. The sills rested on fossil rock.

Before the Dutch came, the land was cultivated by the Mahican Indians, who were later driven east by the Mohawks. The Dutch purchased the land from the Mohawks, replacing corn with wheat.

126

David Babcock is up, left, with Clayton, while Todd Thromback, left, and Henry Suyden lower a long plate from the Dutch barn.

As at Sleepy Hollow, the Van Rensselaer tenants brought their wheat from the barn to the gristmill. Any left over after paying the rent was the leaseholder's to sell or use as he saw fit. Male and female black slaves then worked the land. The founder admitted having fathered a child with a slave named Gin.

In 1917 new owners converted the barn to housing chickens for the Whitbeck Chicken Farm, which included building three additional floors between the original and the roof. The removal of that floor system, some of it during a freezing rain that accompanied a brief thaw, cleared the way to take down the old timbers.

1

2

3

1. David Babcock applies the "gentle persuader" to separate the anchor beam from its post in the Dutch barn.

2. Margaret Babcock helps out by scrubbing off the anchor beam on an uncharacteristically mild Saturday.

3. The crew was surprised to discover the threshing floor virtually intact under a concrete floor that had been poured for chickens. The corrugated metal protected the wood.

4. The Philipsburg Manor crew with the anchor beams disassembled from their section, ready for loading.

5. The crew unloaded the Philipsburg Manor beams by hand in order to avoid marring the patina.

Although we may be aghast at the sacrilege of filling the interior of such a magnificent structure with chicken coops, Richard notes that the conversion may have saved the barn. If it had not been put to a useful purpose, it might not have been maintained. If it had not been maintained, it might have disintegrated.

After one warm weekend when Richard's youngest daughter, Margaret, visited the site to help scrub the dirt off an anchor beam, the cold settled in again, dropping to 30 below zero. Because this was a major project and historical authenticity was required, Richard's crew rigged a gin pole to take down first the rafters, next the purlins that supported them, then the 50-foot-long plates. The long braces, characteristic of Dutch barns, were tied off while the plates were lowered, to prevent their breaking. The crew used the gin pole to lower the anchor beams and their posts, just the opposite of the way the barn was erected. By installing a pulley at the base of the gin pole, old timers and Richard were able to keep the men handling the rope out from underneath the beam they were lowering, an important insurance policy.

As each section was lowered to the ground, the wooden pins were pounded out and the components separated, to be trucked first to the shop at Hancock and then to Sleepy Hollow. Removing a tenon from a mortise requires a pry bar and a large wooden hammer, the descendant of Gramp's "Gentle Persuader." The crew carefully loaded the parts onto a flatbed truck, trying to avoid scratching them or disturbing the patina.

John Harbour from Sleepy Hollow visited the site. During disassembly Richard notified the historical societies of Guilderland and Altamont, so that they could view the process, although the weather did not favor that enterprise. Later he showed them slides of photographs he had taken. Some members went down to Philipsburg Manor to see the barn restored, complete with wooden hinges, cow stalls, horse stalls, wooden eave troughs, and hand-split shingles.

The last step was to jackhammer up the main concrete floor, hoping to find out how the original floor was built. While they expected to find only the slightest remains, the workmen who poured the modern floor had placed corrugated metal roofing over the wood floor. The metal allowed air to circulate

and, when removed, revealed a hand-hewn threshing floor that could be disassembled for reuse. Some of the **Whittling pegs for the Dutch barn at Philipsburg Manor.** supporting beams were also in good enough condition to be reused. The dry stone wall foundation allowed the air to circulate underneath the building, which preserved the wood.

The Babcock crew worked over the timbers at the shop for a month and a half.

Beginning in early April at the new site we began the first step in the erection of the seventeenth-century Dutch barn. We dug a three-foot-deep trench. We filled the trench with small stones packed together for drainage. The ditch continued from the site to a low point to drain away any water that might freeze the ground on which the foundation stone was laid. The stone piers that support the floor system and center columns were similarly ditched. The ramps leading to the various doorways of the barn were lined with retaining walls of stone. We utilized much of the stone from the original site, incorporating about an equal amount from the Philipsburg Manor area. We bonded the stone by overlapping it, not using any mortar, allowing the air to pass through as in the original construction. Springtime being with us, we had a great beginning.

The apple blossoms came out as the crew laid up the stone foundation and placed the sills on top.

We placed flat stones for the footings and the wall, which could rise and fall with the weather. The footing stone was laid approximately a foot wider than the wall above. The upper stone started approximately six inches in from the outside of the footing stone.

I squared the corners as my grandfather had taught me, driving three stakes into the ground at the approximate corners. I ran two parallel lines for the sides, and another line between the two for one end. Then I measured 50-feet down one line, running the fourth side across from that point to the other line. We measured six feet from one corner and eight feet the other way, marking those points. Then I maneuvered one line until the distance between those two points was 10 feet, "which squares the corner," as the old timers said.

Gramp said, "Once that corner is established, all that is necessary is to move in or out the opposite end line to the required

distance and you're home free." As we laid the sills on top of the stone, spring was truly

The sills were carried in by hand and laid in place.

in the air, with the scent of apple blossoms all around us. This is the time when everyone is glad to be alive. As Gramp used to say, "The work is always lighter when the weather is good."

Harbour's plan included filming construction, with the workers wearing period costumes—knickers, knee-high white socks, and loose-fitting shirts. As the weather warmed, the shirts came off. The crew carried beams in the old style, resting on wooden poles which a pair carried. Shiplap joints at the corners of the sill were pegged together. The center sills had a small shoulder to

133

1. David Babcock, left, watches as the Philipsburg Manor smith heats up hand-wrought nails, salvaged when the barn was dismantled, in order to hammer them back into shape. They were re-used in constructing doors.

2. The crew turns the bull wheel to raise the gin pole with the aid of a "raising pole," a forked pole used as a lever.

3. The crew uses the bull wheel and gin pole to raise the third bent, toward the rear of the barn.

4. Tying off the long braces prior to raising the purlin plate beams.

3

4

catch the end of the floor planks, each about two and a half inches thick and 16 to 18 inches wide. They were pinned with wooden pegs. Each plank in the threshing floor was grooved so a half-inch spline of wood could be inserted between two, as a way of keeping the grain from sifting through.

Henry and Richard spent a week constructing a bull wheel from a central shaft they found in a barn in Schoharie. Similar to a capstan on a ship, it permitted two men to do the work of ten. They attached a pole at right angles to the base of the gin pole as a "raising pole." By pulling a line from it through a pulley and taking up the slack with a bull wheel, they sent the gin pole skyward.

By this time, a crowd was begining to gather each day to watch progress. One mother brought her two children regularly.

The guy lines were tied off. The crew took out the raising pole and rigged the line through a pulley at the base of the gin pole. In the early days, slaves would have turned the bull wheel.

The crew inserted raising pins of oak into the bents that were about to be airborn. Across these pins planks would be laid to create a work platform. With the bull wheel pinned to the floor and the gin pole positioned, the men hauled on the first bent, made up of an anchor beam and two posts, while Henry and Richard guided the tenons in their bases into the receiving mortises. "The bull wheel squeaked along with the pulley blocks," Richard later wrote, "as the section of timber rose slowly and majestically to regain its rightful place in today's living history." Spectators cheered.

Thus the crew propped against the sky and plumbed three anchor beams. Richard believes this was the first authentic raising of a Dutch barn in this century.

As the bents went up, planks were tacked on and held them in place temporarily until the horizontal girts were inserted to lock them in place. The girts were heavy and the anchor beams had to be pounded to get the tenons in, which meant the men holding the unattached ends of the girts took a beating. The men occupied any spare time straddling the shaving bench to carve pins from freshly cut oak. These were kept in a small bucket that could be hoisted to whomever was pinning timbers together. Philipsburg Manor's smith straightened the nails Richard had salvaged at Guilderland.

A chance remark led to a demonstration of agility. Richard commented that the top of a nearby pine would present a great angle for photographing the barn. Before he knew it, David had starting shinnying up the first 30 feet, which were bare of branches. Not to be outdone, Clayton followed. They did manage to photograph the site.

The crew moved the gin pole, which was repositioned for each bent, to hoist the plate beam, which would link all the anchor beams at their tops. The gin pole should never be more than three feet away from the resting place of the beam it is raising, or it may bend and break. The long sway braces were tied in place to receive the plate.

Next the crew erected the smaller framing that builds out a Dutch barn on each side of the anchor beam. Then, after three weeks of work on site, they rigged the gin pole with a jib pole to raise the long rafters. The jib pole raises the rafter from the ground outside the structure. While one man balanced it over the outside plate, another hitched it to the gin for the final lift. Once on the plate and purlin, the rafters were rolled into position. The effort required knowledge and team work.

The rafters pinned to purlin and plate, roof boards were nailed in place, spaced to allow air to dry the shake shingles, split from cedar, from the inside. Treated this way, without any roofing paper, shingles will last some 75 years, Richard says. Dove holes under the peak helped provide ventilation.

The crew constructed wooden scaffolding, as in the old days, from which to roof. "You have to know to drive your nails downhill a bit," Richard says, "and also to avoid picking out a board with a knot. My grandfather used to stretch a plank across a couple of beams on the ground spaced as far apart as the supports would be in the air. Then he'd jump on it in the center to make sure it didn't break."

They had to replace the siding with new 1-by-12-inch boards, for the siding they found dated only to 1917. The siding was nailed in place horizontally. In the years since 1982, it has weathered so that it looks in keeping. Richard spent a week with a gutter adze hollowing a 50-foot timber to use as an eaves trough to catch the water coming off one side of the roof. Two-inch-diameter pins,

The Philipsburg barn is now covered with spaced roofing
boards, which will allow the wood shingles to breathe.
Below, Jack A. Sobon drew this section and details of the
Philipsburg barn.

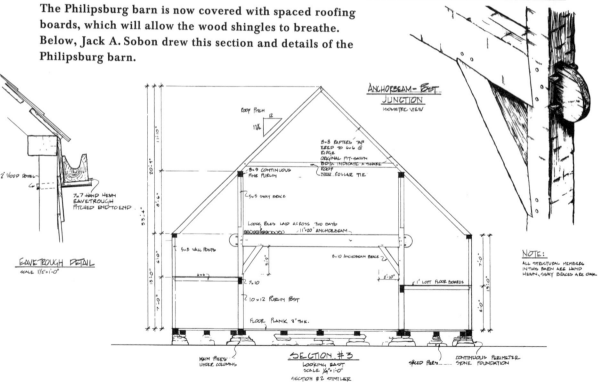

inserted in the original holes in the plates, held the trough. Richard's son, Clayton, did the backbreaking work

The Philipsburg barn boarded in, complete with dove holes and pentice roof over the doors. For sheathing the crew had to use newly sawn boards, which soon aged to blend with the original timbers.

of hollowing the companion trough. Those gutters have had to be replaced.

The crew laid poles across the anchor beams to hold the wheat. They rebuilt the hayracks. Draft horses, a Morgan mare, a Belgian stallion, and several cows were purchased to replace the ones lost in the fire. About a dozen Wiltshire-Dorset sheep, including five that survived the fire, were no longer kept in the barn but in a nearby sheep cote. Oxen are currently kept in the barn.

Sleepy Hollow, endowed by John D. Rockefeller, Jr., runs 25 acres now, including manor house, gristmill, and barn. As the director of public information, Nancy Gold, points out, the substitution of the Oxburger barn actually improved the authenticity of the operation, since it more accurately represented the period of the estate than the one that burned. "It was the only job I ever made any money on," Richard says. Although at $250,000 it was Richard's best paying job, the money was more than earned.

The newly rebuilt Philipsburg manor barn (right) in its setting beside the old mill. *Opposite:* **Richard spent "a long time" hewing out an eaves gutter with a gutter adze.**

The Wemp Barn

In the 1990s the next generation of Babcocks saved another fine specimen of a barn, built on the same scale as the Saratoga barn and the one at Philipsburg Manor.

A barn built for Jan Wemp in the early 1700s in Fort Hunter, New York, west of Schenectady, was up for sale and had a potential buyer, who intended to move it to Montana. The late Vincent Shaffer, president of the Dutch Barn Preservation Society at the time, became concerned at the possible loss of an important piece of New York State's heritage. He contacted a friend, Albany businessman Carl E. Touhey, and convinced him to pay to move the barn to his land on Onesquethaw Creek Road in Feura Bush, just south of Albany. Touhey contacted a number of contractors before settling on Babcock Barns.

Richard arranged for his son Clayton to take charge, with Charley's help. They disassembled the Wemp barn carefully, saving every board and nail they could, so that the reassembly in Feura Bush was as authentic as possible. Richard helped out by reconstructing wooden hinges and the hayrack. The Babcocks reinstalled the 1850 slate roof which, although not part of the original barn, will protect it for decades to come.

Touhey lives in the 1755 stone Garret Van Zandt house on the property. So persuaded was he of the value of the barn that he placed it in a trust to assure it would never be taken down and could be available for public view during daylight hours.

IX

NORUMBEGA

AS WELL AS LEARNING the history of barns, Richard has studied history through barns. In ways not always emphasized in textbooks, he has learned how things were by studying the artifact itself. For example, he learned from the types of hewn timbers that Germans as well as Dutch were settlers in the Hudson and Mohawk valleys during colonial times. He learned from the sheer size of the buildings, as well as through deeds and other documents, that large farms in New York and New England flourished on the bent backs of slave labor.

With Richard we can step across the threshold of yet another barn to speculate on something that appears in few history books: the possibility of sixteenth-century French settlement in the Hoosick Valley of what is now New York State. This particular barn stood in Hoosick, south and east of Hoosick Falls, New York. Richard acquired it in 1983 and disassembled it in 1984. Because its timbers and the proof they offered of an extraordinary antiquity were stored in the Great Plantation Barn of Saratoga, they burned in the fire of July 1992.

I must have seen this barn before, having driven Breece Hollow Road many times on my way to Stephentown from Bennington. It is a shortcut from Route 9 out of Bennington, Vermont, to routes 376 and 22, which I must have learned shortly after getting my driver's license. That year I worked on the Robert Fowler farm out that way.

This time my eyes were opened. Oh, yes! How many times we all go down roads not seeing something that could make a real difference in our lives. Prior to this day I was blinded by the knowledge of where first settlements were supposed to be. I had learned from old histories to look for barns only where first settlers settled, not seeing the barn for the history.

This day what caught my eye was that steep roof, with a hole in it. Why not take a look, I figured. I guessed the owner would want to sell. I stopped at the farmhouse. The older Walter Reitz, the man who had worked the farm for his lifetime, and his son, Walter, who seemed to be cut from the same cloth, were interested in my seeing the barn. Young Walter gave me the tour, his father following along behind. "I've been meaning to call you," the younger man said.

He told me a man visited the barn a year before who wanted to move it to Cape Cod. "I had heard of you," he said, "and I wanted you to see it first before I let it go."

When I entered that barn through the folding doors on the south, I had 30 years of experience, research, and understanding to see it with. That barn drew me. It put me to work for its benefit.

Inside the barn for the first time with the younger Walter, I was surprised. It was different from any other I'd seen. The color of the beams or patina was deeper than any I'd seen before, running dark brown to gray.

The first architectural element to catch my eye was a great swing beam, in a section of beams to the right of the entrance doors. It ran across from post to post above my head without a center post or brace. Quite a span. Above it, another beam ran parallel. This one was smaller, with a brace at each end. Above that beam stood two purlin posts, bearing the weight of the roof. Smack dab in the middle of the great beam was a king post—something I'd never seen before in a barn.

The king post was joined to the great beam with a half dovetail tenon. A locking wedge was driven in on one side. In the center of the dovetail was an oak peg that passed through the entire beam. At one time the king post rose to the roof, but had been cut off above the smaller cross beam. No doubt that step was taken to make room for a hay fork. Although the amputation weakened the structure, at least the bottom of the king post was left, to help in visualizing the original construction. For a king post to work, it has to have long supports from the head to carry the weight to the outer wall posts. They had been removed down to the purlin posts, as part of the hay fork operation.

I knew I could re-create the barn from the truncated parts, which is what I love to do. I've done it many times.

Next I noticed marks of the carpenter's compass on just about every beam. I had never seen so many left before—not for show but strictly for business.

The French barn after the nineteenth-century metal roof had been removed, with Mount Anthony in the background.

Each corner post had two great braces from the tops of the posts to the sills, used to keep the wind from wrenching the frame. These great braces are not found in timber-frame barn construction of the seventeenth and eighteenth centuries. They intersected with half-lap joints the two outside wall girts—smaller cross beams between wall posts used to nail on boarding. They were in addition to the regular braces at the intersection of cross beams with posts throughout the frame. Between them they appeared to brace a fortress against wind and tide. Each brace was scored, hewn, framed, and numbered. The numbering was left chiseled into each member in Roman numerals. The feeling they gave was sheer business: built with no fooling around.

An unusual feature of this barn ➤ is the braces that extend from the plates to the sills, with a half-lap joint where they cross.

Richard bought the barn for his museum, not as part of his stock. He patched it up temporarily, including the hole in the roof, having to wait a year before he had money and time to move it. In between finding the barn and taking it apart, he scoured the Registry of Deeds and Surrogate Court for the section of the Van Rensselaer Patent that related to the barn. He could trace the records back to 1767, but he knew the barn must be older than that. This barn held many mysteries; a true challenge for the Sherlock Holmes of barns.

Dismantling and Discovery

In the summer of 1984 he, his son Ronald and Jim Haley, whom he hired for the job, began the long task of taking it apart, using a gin pole. Richard notes that "many times when we were ready to lower a beam, the Reitzes would help, leaving their farm chores, both old Walter and young Walter."

When Richard and his relatives began to pull off the roof boards, the first step in taking down a barn, they found a set of 13 concentric circles, the largest eight inches in diameter, with two smaller circles, etched by a carpenter's

compass in the ridge beam of the barn. The two smaller circles, located at four and eight if the entire design were a clock, had their centers on the outer ring and were large enough to encompass five of the thirteen circles each. Part of the marking had been eaten away by time and weather.

The only person whom the builder could have expected to see it would be someone taking down or repairing the barn. Richard believes that master builders, knowing that farms change, built barns in anticipation that they would be disassembled to be moved. He is familiar with the swings of the compass left as a guide to the dimensions of the building. He saw no way, however, for this design to work as a plan.

One of the intriguing pleasures of taking apart old barns is to receive messages from the original builder. For example, Richard has often found crudely sketched large heads on barn planks. Thus the master builder represents himself—the head large enough to hold the brains of the operation.

Because the roof boards on the Reitz barn seemed to be original, Richard guesses he was the first to find these circles since they were drawn. He wondered: what was the message? He contacted James Corsaro of the Albany Archives, who searched in vain for anything like the circles symbol.

Usually braces extend a few feet where posts and beams join, as branches might near the top of a tree. Richard has not found braces extending from the tops of the posts to the sill in any other barn. His studies indicate they may have been common in Europe in the fifteenth and sixteenth centuries, however. More sophisticated builders found them unnecessary, yet here they may have played a role in keeping the structure standing for a long time. Such long braces might be related to *saltire*, a French style of two long braces that crossed in the middle with half-lap joints. The design derives from the St. Andrew's cross of equal-length members, used in heraldry.

The nails in this barn were cruder than even the earliest handmade ones he had previously come across.

Old timers marked structural members that joined together with paired Roman numerals, such as I, II, III…IX, X, because that form of numeral was easier to strike with a chisel than was the rounder Arabic. In this barn, however,

three ninth (IX) joints were marked with arrows. Richard has never

The mysterious 13 circles, with the two smaller circles, about eight inches in diameter, on the ridge beam.

seen that elsewhere, but understands arrows were used in early French construction to designate the number nine.

Richard came to the conclusion that the barn was built very early by an ethnic group whose work he had not come across before. That ruled out German, Dutch, English, and the related Scottish and Irish. To find out which group might have produced the barn he plunged into local history. If books on national history couldn't help, perhaps he could find a clue in the traditions and stories of the area.

Local History

In Grace Greylock Niles' *The Hoosac Valley: Its Legends and Its History,* published in 1912, he read about the American Indian legends of St. Croix, a "spirit's cross" formed by the T-shaped junction of the Hoosic and Walloomsac rivers, near the barn site. Niles said that when the Dutch arrived in the area they found the ruins of a fort at that junction—and that they, and later the English, built

151

Arrows were used instead of the Roman numeral IX.

forts on the same site. A historic marker on New York Route 22 indicates the plateau—currently a cornfield and a gravel quarry—on which the forts were built.

Giovanni da Verrazano, an Italian explorer in the service of France, described hearing about a magnificent city named Norumbega on a great river south of Cape Breton in 1524. Niles wrote that Jean Alphonce's crew from St. Ange, France, visiting the Hoosac Indian hunting grounds between 1540 and 1542. Niles said they named a site St. Croix and established a modest community there. Alphonce said he found Norumbega; he did not mention founding a settlement.

Niles wrote that a Jesuit in Alphonce's group blessed the mingling waters and raised the banner of the Holy Cross on a high terrace overlooking the valley. "The traders later built the palisaded castle St. Croix," she wrote, "and founded a forest chapel in memory of the missionary, St. Antoine of Padua."

When Grace Niles wrote about Jean Alphonce, she was a detective with a hunch. She was convinced that Alphonce started that St. Croix settlement. She knew the legends of St. Croix and the forest chapel, as had her sources, Arthur Latham Perry's history of Williamstown and Sylvester's history of the New York area. She did not read Alphonce's actual writings. Once you do, you can more easily place him at the head of navigation, Albany and Troy, in 1542.

Perry ponders why the French name St. Croix, spelled in different ways, should cling to the area where the Battle of Bennington was fought, as it still does. "It is a cherished opinion of the writer," commented Perry, "and one originally suggested by personal observation at the two localities concerned, that the name 'St. Croix' was given by the French missionaries to the district to account for the fact that the Walloomsac strikes the Hoosic exactly at right angles..." and he writes of an "ancient church of St. Croix" that disappeared before 1900. It was located near the river junction, he said.

The fort site, surrounded by the two rivers and wetlands, is on a plateau, a large, flat area, with an embankment of about fifty feet running down to the river on the south side and another embankment of about twenty feet extending to the river on the west side. Easy to defend in the event of attack, the site would have been attractive to a small exploring party seeking security.

Niles noted that just as the name St. Croix still clung at the beginning of the twentieth century to that section of the Hoosic, St. Antoine clings "to Mount St. Anthony of Bennington, in whose shadow St. Antoine's chapel undoubtedly stood." The name predates English and Dutch settlement in the Bennington area. The Hoosacs and Mahicans had passed the story down, so that they welcomed Henry Hudson and his Dutch crew of the *Half Moon* in 1609 as if Alphonce were returned at long last from St. Ange.

To Richard it was as though he had put in a long-distance call to Grace Niles. She had listened carefully to what he wanted to know. And then she answered his question. The Reitz barn was very early, French and originally a chapel.

Somehow, based on what I had discovered in the architecture of the Reitz barn, I believed her.

Soon after, I asked young Walter, "Where is Mount Anthony from here?" He said, "Right there," as he pointed to the highest peak to the east of the barn.

I said, "You're kidding me."

"Nope, that's it."

I told Walter right there that this Grace Niles who wrote a history of the Hoosac Valley told of a legend of a forest chapel being built in the shadow of Mount Anthony. I had the feeling, immediately, that this barn could be it. Walter did, too, and assumed that it was discovered by later settlers and reused as a barn. King posts were common in churches. Barns and churches have always been related. The architecture was sure agreeing with it as far as I was concerned.

Figuring I had made a big discovery, and being no foreigner to the press, I informed the local newspaper about my thoughts of the barn being the chapel. Well, that was just great and I took a glory ride on the press for a while, including Channel 10 television in Albany. Eventually, though, this led to an article in the *Berkshire Eagle* of Pittsfield, Massachusetts. The reporter, Daniel T. Keating, sought comments from professors of history at various colleges like Harvard and Williams. They disagreed, saying Grace Niles was not reliable.

Boy! Was I shot dead in the water. Of course, I argued on the basis of the structure. I was basically done for, for the time being.

Actually, Sung Dok Kim, chairman of the history department at the State University of New York at Albany, said: "Oh, no, it's pooh-pooh. There were no settlements in this area in the sixteenth century. It's totally out of the question." Abbott Cummings, professor of American decorative art at Yale, said Richard's claim was "probably preposterous." He said sixteenth-century explorers, such as John Cabot and Giovanni da Verrazano, did not get as far inland as Albany.

Historians generally have agreed that the French did not settle in the area before Champlain arrived in the St. Lawrence Valley, or Hudson arrived on the river that bears his name, both in 1609. The easy truth has always been that the French tended not to settle anyway, but to go out to trade with the Native Americans as *les couriers de bois*—the runners of the woods. The British and the Dutch, on the other hand, established villages and planted crops. The French strategy appeared at first to be better for making friends with the native peoples, but ultimately it lost them the continent.

Most historians have never heard of Grace Greylock Niles. They are familiar with local history as an often charming but "iffy" proposition. Those who know of Niles stress the title of her work, pointing out that she mixed history, which can usually be supported by documents, with stories or legends, an oral tradition. Generally she doesn't differentiate. Williams College Professor of History Emeritus Frederick Rudolph called Niles "very unreliable" in the *Eagle* article.

Another historian suggested that if the barn were indeed of French construction, it might have been built by French Protestants who came south from what was to become Canada just before or during the French and Indian War. They did not wish to remain in a Catholic country. In fact, this theory conforms to the date to which Richard could trace the barn. This historian might have reminded Richard what he himself has proved so often, that barn building is a traditional skill. For example, Huguenots from Nova Scotia might have continued for generations to build barns as their ancestors did, being cut off from innovations that evolved in Europe.

Some of these Acadians who sided with the British might have migrated down Lake Champlain and Lake George, or sailed through the Gulf of Maine into Long Island Sound and up the Hudson in the 1760s. A barn builder among them could have brought with him the techniques—and possibly even the nails—that his great-great-great grandfather used in the old country. That, of course, would still make the Reitz barn a tremendous find. Richard knows of no other roots French barns in New York or New England. Even in the Province of Quebec, most of the extant earliest barns are German or English. Richard feels certain that if he could go to France, he could find some compelling physical evidence that would tie his barn to French ancestry. He quotes Perry as saying the name St. Croix could only have come from "old Catholic."

Richard's immediate response to the skeptics was to get the barn down and stored. Still, he believes in his hunches. He read Niles again, more carefully, and began to do exactly what a good historian would do: read her sources, even the originals of some sources she used at second-hand. Richard, who had left Williamstown High School at age 15 to plant utility poles, journeyed 130 miles across the state to the Boston Public Library. It did not have Alphonce's log, but the librarian gave Richard a letter to the Harvard University libraries and the advice to take the subway there rather than drive, as parking was difficult in Cambridge.

Both worked. Once at Harvard he and his son, Charley, were shown seats in the rare books library. Then Alphonce's *La Cosmographie* was placed before them.

"It was truly an honor just to be allowed to see it," Richard says. Unable to read French, he looked through it for the word "Norumbega," the fabled city Alphonce said he found. Richard asked the librarian to copy each page that held that word. The photocopies arrived in the mail a few days later. "A miracle of sorts," he says. He had the pages translated.

Myth and History

Among early explorers, Norumbega appears to have been a myth, transferred from Mexico to a more northerly clime, of a city paved with gold. Richard believes this place, real or imaginary, was located not on the Penobscot, as Samuel de Champlain supposed, but on the Hudson near the Hoosick Valley—or that it was a general place name that included much of what is now New England. He believes that Norumbega and St. Croix were related.

In the first decades of the sixteenth century, Hernando Cortes began to loot the treasure of Mexico, setting off a European gold rush to the New World. France challenged Spain by piracy, capturing vessels laden with gold. One of these French seamen looking for a way west, Jacques Cartier, sailed north of Newfoundland and into the St. Lawrence in 1534, and all the way upriver to Quebec and Montreal in 1535. Both times he brought kidnapped American Indians back to France with him.

In 1541, he tried colonization, building a fort on a point above Quebec.

Because Sieur de Roberval, with three ships holding 200 settlers, was delayed, this first attempt by Europeans at true settlement in North America failed.

One of Roberval's pilots, Jean Alphonce of St. Ange, reported sailing southward in 1542 to a great bay in the latitude of 42 degrees; he then continued "to Norumbega." Where did he go? Henry F. Howe, in his *Prologue to New England*, describes the expedition to Massachusetts Bay, and lauds the sailors for their ability to spend five months crossing the Atlantic and still be able to hit an intended landfall in the New World. If Alphonce said 42 degrees, he was probably close to it.

Vague references to Norumbega began to appear, attached to a great river somewhere between Cape Breton and Florida. Andre Thevet wrote of his adventures after returning from his 1556 voyage. He said he arrived at a river that the Indians said Frenchmen, who lived upstream, called Norumbega. He described the area as follows:

> Having left La Florida on the left hand, with all its islands, gulfs, and capes, a river presents itself, which is one of the finest rivers in the whole world, which we call "Norumbegue," and the aborigines "Agoncy," and which is marked on some marine charts as the Grand River. Several other beautiful rivers enter into it, and upon its banks the French formerly erected a little fort about 10 or 12 leagues from its mouth, which was surrounded by fresh water, and this place was named Fort of Norumbegue.
>
> Some pilots would make me believe that this country is the proper country of Canada, but I told them that this was far from the truth, since this country lies in 43 N. and that of Canada is 50 or 52. Before you enter the said river appears an island surrounded by eight very small islets, which are near to the country of the green mountains and to the Cape of the islets. From there you sail all along into the mouth of the river, which is dangerous from the great number of thick and high rocks; and its entrance is wonderfully large. About three leagues into the river, an island presents itself to you, that may have four leagues in circumference, inhabited only by some fishermen and birds of different sorts, which island they call "Aiayascon," because it has the form of a man's arm, which they call so. Its greatest length is from north to south. It would be very easy to plant on this island, and build a fortress on it to keep in check the whole surrounding country.

Richard found Thevet's *La Cosmographie Universelle* at Harvard, as well as Alphonce's *Cosmographie*. Howe, following Champlain, makes the case that this description neatly fits the mouth of the Penobscot, in Maine. Richard, following regional historians, believes it describes a different river. By this time French traders, perhaps including Alphonce and Thevet, appear to have made their way up the Hudson, even though it was yet to be "discovered" by Henry Hudson. To Richard, Thevet's or Alphonce's account is an apt description of leaving Massachusetts Bay, following west through Long Island Sound, through Hell's Gate, and up the Hudson. He supports that view with reference to contemporary maps, approximate recorded distances, and the latitude mentioned—whether 42 or 43 degrees.

The most famous map was that of Gerald Mercator, in 1569, which indeed shows a river placed about like the Hudson, branching at what is called "Norumbega." The left branch could be the Mohawk. A right branch could be the Hoosic. Mercator's main source was Jacques Cartier, who may have gotten his information from Alphonce.

Further evidence exists in Richard Hakluyt's account of David Ingram's extraordinary 1569 hike from the Gulf of Mexico to find a Christian captain to take him home—or to France, as it turned out. (Hakluyt includes Ingram's narrative in his first history of "The Principall Navigations, Voyages, and Traffiques and Discoveries of the English Nation," printed in 1589, though he omits it from his two later volumes, saying only that it contained too many "incredibilities.") Richard, reasoning from Ingram's account, concludes that Ingram ascended the Hudson and the Mohawk, then passed around a great sea and out the St. Lawrence River Valley. On the way, Ingram came across "Bega, a country and towne of that name, three quarters of a mile long, where are some great stores of dry hides." The "Bega" referred to could certainly be the same as Norumbega. Alphonce had also mentioned the stores of hides at his Norumbega—such as would be appropriate to a traditional American Indian trading site at the junction of the Mohawk and Hudson rivers.

Ingram continued on. "The said Ingram," Hakluyt wrote, "traveling towards the North, found the main sea upon the north-side of America, and traveled in

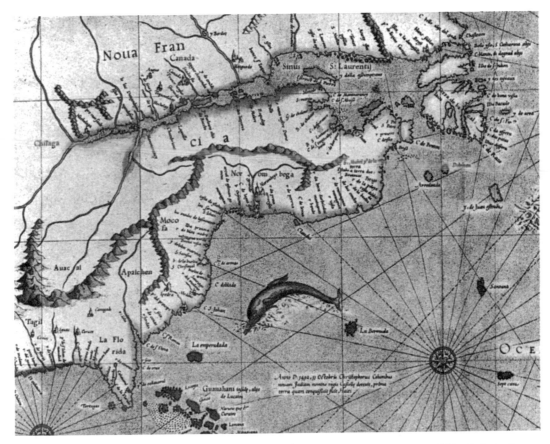

the sight thereof two whole days." Richard believes the sea to have been Lake Ontario. As he points out, it would be hard to make sense of the reference if Ingram had ventured

Mercator's 1569 map of North America shows a river positioned like the Hudson, splitting in an area called "Norombega." (Used by permission of Chapin Library, Williams College; Ernest LeClair photo)

up the Penobscot. Ingram also recounted visiting a town he named as "Sagunah, a towne almost a mile in length." Ingram then made his way to Ochelaga, an American Indian town on the St. Lawrence, site of the present-day Montreal. Ingram's mention of Sagunah and Ochelaga conforms to names identified in Cartier's account and seems to pin down the route of his travel.

"After long travaile," the account continued, "the aforesaid David Ingram with his two companions Broune and Twide came to the head of a river called Garinda, which is 60 Leagues West from Cape Breton, and there found a French captain, Monsieur Champagne, who took them in his Shippe and brought them unto Newhaven, from thence they were transported into England, Anno. Dom. 1569."

Ingram describes a great rock of crystal at the head of mighty waters. "Head" could refer to the end of the navigable portion, or what was to become Albany. A. J. Weise in his *Discoveries of America to 1524* defines *Norumbega* as a corruption of old French words meaning something like "steep wall of stone that borders a river." Weise points out that Alphonce's statement that the river Norumbega was "salty at more than 40 leagues inland" could fit "the Hudson and no other." Weise thought Norumbega referred to the Palisades.

But Richard read New York historian Nathaniel Bartlett Sylvester's story of Diamond Rock, across the Hudson from the entrance of the Mohawk. Here Moneca, an Indian mother, kept a fire burning many years for the return of her sons. The old Indian who told the story said that here, later, the white man used the rock as though it were a beacon.

Richard scoured the east bank of the Hudson for such a rock—and found it, just at the upper extremity of the navigable portion of the river, at 42 degrees latitude. Today that outcropping is a landmark several hundred feet above the Hudson and 30 feet above a condominium housing development at Lansing-burg, just north of Troy. Richard is the first to suggest a relationship between Diamond Rock and Norumbega.

I found Diamond Rock in an abandoned field, with young trees growing everywhere. Although it is not covered with crystals, its present name comes from the sun-reflecting sheen of its surface. I climbed up and stood on top. I saw it all from up there—if anything will convince anyone, being there on that rock will do it.

Alphonce tells us in his cosmography that he believes his river would run into "Ochelaga" or the St. Lawrence. On no other river would that be a reasonable inference to make when one was at 42 degrees. Although the Hudson doesn't quite make it to the St. Lawrence, it points in that direction.

Diamond Rock, photographed before a housing development was built beside it.

I could see down the river to Albany, to the Mohawk tumbling over the falls to join the Hudson, and far up the Mohawk Valley. Directly below I could see several islands at the river's head of navigation. To the north I could see it seem to disappear at Stillwater, where the Hoosic joins.

I felt Norumbega in my bones. This was my greatest moment of conviction. I was standing on the reason for the name, given so naturally years ago. I was one happy guy, convinced at that moment that I could write the story that would convince all.

Norumbega was the entire area about the navigable head of the Hudson and on up the Hoosic to my old barn. The Hudson itself would have borne the name—and perhaps New England all the way to Cape Cod.

Niles, following New York historian Cuyler Reynolds, believed that Alphonce or—in another part of her book—a Jesuit from Canada, started a settlement at St. Croix, which led to a fort and a forest chapel. Richard notes that she did not read Alphonce, which may be why she equivocates, suggesting the possibility of a separate Jesuit expedition. Richard, characteristically, therefore read the Canadian Jesuit records, finding no mention of establishing a settlement on the Hoosic.

On Alphonce's ship were carpenters, masons, farmers and priests, men and women intending to form a colony for Cartier in the Quebec area. While Alphonce does not mention leaving anyone behind when he reached Norumbega and decided to establish St. Croix, these passengers may have settled in St. Croix instead of going on to Quebec. Champlain so believed that people from the Cartier expedition were left behind that in 1602 he searched for survivors of the Norumbega settlement. Where he looked, without success, was the Penobscot, which to him matched the physical description. Perhaps if he had looked on the Hudson, Richard says, he would have found the Cartier party's descendants.

Or perhaps not. Niles writes that in 1590 the Mohawks drove the Mahicans out of the Hudson Valley and up the Hoosic River. Perhaps Fort St. Croix was overrun at that time. Richard maintains that the barn was the settlement's original forest chapel. The fort and the barn were about 10 miles apart, so that the destruction of one would not necessarily have led to the destruction of the other. The chapel might have remained, a curious antique in the wilderness, until a Dutch settler found it, some time after 1620, and converted it to a barn.

A Blessed Barn

"Even though I sometimes have doubts about a building lasting so long," Richard says, "my barn was blessed." He can suggest a series of circumstances that preserved this barn almost 450 years. The fact that someone found a use for it meant that someone would have kept it up. The steep pitch of the roof would have shed water, even when many shingles were rotted. The fact that it was built where no town grew would have kept it from being cannibalized for other build-

ings. The dry stone wall foundation would have permitted the circulation of air so important to

The French barn being disassembled at Hoosick, New York.

maintaining wood. The fact that it was overbuilt, as shown by those long braces, made it sturdy enough to withstand weather and shifting ground.

Furthermore, Providence must have smiled on it. Six years before Richard found it, an adjacent, more recent barn burned to the ground. The local fire chief expressed surprise that the old barn survived. And, one cannot discount Richard's own intervention, at a time when a hole in the roof threatened to rot the frame and the north wall, at least, was pretty well gone.

And then, those mysterious concentric circles spin before our eyes. What is their message? Could they have a religious significance, as would be appropriate to a wilderness chapel? Perhaps they were scribed not only for a builder to find but for the Builder who sees all. Christ was a carpenter, born—at least in Northern Renaissance art—in a barn. In the great cathedrals of Europe the craftsmanship and detail, even the ornamentation, exist high up, out of sight of any human eyes.

Reverend John Eusden, for whom Richard built a barn house, suggested the circles might represent Moses and the two stone tablets containing the law; or Christ, in the center, and the 12 disciples, with Moses and Elijah as the smaller circles. One can imagine a lonely builder, remote from the world of his ancestors, scribing his statement of faith into the building he hoped would save his soul.

This barn appears to embody the history of any roots barn still with us, even though older by 150 years—the key factors being strength, good workmanship, maintenance, caring, and Providence. Sadly, fire did destroy it at last —another warning, if any more were needed, of the precarious state of historic barns. Yet fire did not destroy the main beams which, although charred, still exist as evidence.

History as Hypothesis

The location of this barn alerts us that not all early settlements were where they were supposed to be, along the great rivers: the Hudson, the Mohawk, the Connecticut, the Charles. Some areas sufficiently settled to require a barn or even a chapel were in remote and idiosyncratic places. Even the possibility of sixteenth-century French settlement reminds us that just because historians have written something down doesn't mean it is true. History is often a hypothesis, waiting for further evidence.

Consideration of the period of this barn, like the other roots barns Richard has identified, reminds us that Europeans lived in this country for a longer period before the Revolutionary War than they have since. In fact, we don't really know who were the first settlers or when they arrived. While Plymouth Rock in 1620 is a good foundation stone upon which to rest our democratic experiment, as Howe points out there was a century of European exploration of New York and New England before that...not to mention earlier visits by fishermen of various nationalities, and settlement by Native Americans stretching back into misty, prerecorded time.

Richard's main case for this special barn is argued by the structure and by written sources. It was a unique structure, historically important, of unusual lineage for northeastern North America. To generalize from that, every one of the fast-disappearing ancient barns has its own mysteries and reveals new discoveries to those who take the time to attend to them.

In 1869 John Greenleaf Whittier wrote a poem in which, after a long search, a knight thinks he finds the fabled city at evening. Instead he finds death. His squire buries him, leaving the old and mossy cross that Champlain said he found in the woods on his fruitless search for Norumbega.

...At the shut of day a Christian knight
    Upon his henchman leaned.

The embers of the sunset's fires
    Along the clouds burned down;
"I see," he said, "the domes and spires
    of Norembega town."

"Alack! the domes, O master mine,
    Are golden clouds on high;
Yon spire is but the branchless pine
    That cuts the evening sky."

"Oh, hush and Hark! What sounds are these
    but chants and holy hymns?"
"Thou hear'st the breeze that stirs the trees
    Through all their leafy limbs."

"Is it a chapel bell that fills
    The air with its low tone?"
"Thou hear'st the tinkle of the rills,
    The insect's vesper drone."

"The Christ be praised!—he sets for me
    A blessed cross in sight!"
"Now, nay, 't is but yon blasted tree
    With two gaunt arms outright!"

∝

"Yet onward still to ear and eye
    The baffling marvel calls;
I fain would look before I die
    On Norembega's walls."

—from "Norembega"
by John Greenleaf Whittier

X

PLUMB AND SQUARE

AS WELL AS TAKING DOWN barns and re-erecting them, Richard has repaired the beams and sometimes replaced them. In the last 10 or 15 years he has frequently framed recycled beams into other old barns. When these beams were irregular, he used the scribe rule to guide the work. Replacement beams were generally salvaged from barns in poor shape otherwise or not sufficiently distinguished to save.

Sometimes in the course of his work he felled trees and hewed them to become replacement beams. He did that on the first barn he moved, with Gramp, for example, and other times as well. Sometimes he hewed at events such as the annual Hopkins Forest Fall Festival at Williams College, where he annually demonstrates old skills.

Yet a challenge remained. He had never framed an entire barn by the scribe rule. He had not taken his compass and his line to lay out and cut all the sills, the posts, the plates, the braces, and the rafters. He wanted to erect a small barn, laid out in the old way. He knew it would work, because he had been using the method in his restorations.

When Henry N. Flynt, Jr., chairman of Williamstown's 1753 House Committee, asked him if he would like to participate in the 40th anniversary celebration of the '53 House in September of 1993, Richard realized that this was the perfect opportunity. What about a barn raising? he asked.

Henry Flynt thought raising a barn near the 1753 House at Field Park sounded just right. The 1753 House had been built by volunteers for the town's bicentennial, using the methods and materials used by the first settlers of the town. The ash framing members were felled with a crosscut saw and wedges and carried by hand from the White Oaks section of town, where the original settlers cut oak for their homes. Although the beams were measured with a straight-edge and square, the framing tools were an auger and chisel. The mortise and tenon joints were pegged together. The fireplace was laid up without mortar. Volunteers used a froe to split clapboards and shingles out of straight grain oak. Yes, Richard's was the appropriate barn for that house.

Time did not permit beginning by hewing out the beams, nor was that necessary for him to demonstrate how the scribe rule could be used to frame. He used beams that he had accumulated from old barns. Twelve was the number he chose: his would be a three-bent barn, measuring approximately 12 feet by 12 feet by 7 feet to the plate; even the roof was 12 over 12, a way of saying that for each foot it rose it slanted in a foot.

The dimensions are approximate because Richard measured by the carpenter's compass rather than with a tape. He thinks the setting he used was about a foot, but what he went by was 12 turns when he walked the line with the compass. He could have used any setting. He inscribed an arc on a board. Each day when he returned to the job, he set his compass to the radius of that arc. Inches and feet were not part of the calculations.

This little barn was special to me, since I had framed it all out of old oak timbers, using tools similar to those of 1,000 years ago— a line rubbed with charcoal, a carpenter's compass, a chisel, and a wooden mallet. It is not a rebuilding of an original at all, I simply used old timbers, reframing them to my own specifications. I designed it to be used as art, yes, "barn art."

I designed it to be left open, to be seen from all sides. Once it had a permanent site, I planned for it to have a wooden shake roof, as well as to be fastened to a foundation.

I did not use a ruler, only turns of the compass. Understanding the compass, knowing where and how to install plumb lines on posts and cross timbers, choosing best sides and second best sides, and so on, makes the scribe rule. By it I can square a frame, cut all the timbers at proper length, without using feet or inches.

Think of it. All those old guys had was the compass, yet they built structures plumb and square, and everyone wonders at how well they have stood.

Richard and Henry picked out the best of the timbers they had on hand. They found beams long enough, even after they had cut off the original mortises and tenons and allowed for the new ones. They chose beams that did not have the remains of joints in the sections to be cut. The best of these they set out to be the six posts. Others they selected for sills and plates and cross pieces.

"It's an advantage to have a slight bow in these," Richard says, "because when you arrange them bow side up in the frame, the weight of the building straightens them." In contrast, straight beams might pick up a slight bow downward.

This barn didn't represent any "roots" form. Except for the scale, the design might have been Colonial, picking out certain features from a variety of older structures. That's what early framers did after they had lived in this country for several generations. Richard and Henry laid it out by the scribe rule, the heart of which is seeing the perfect, square timber inside the actual irregular beam. That approach is a lot like judging people.

They chose the best side of the timber as a starting point. The "best side" means the straightest. It will always face out. Framing begins on that side of each beam and works around to the other three sides. At each end of the beam they found the center point by adjusting one point of the compass until the other could make an arc always on wood. Then they swung a half circle, with the compass, at each end so that they could snap a chalk line between the center points. Next, using the compass, they walked off, point on point, along the center line the locations of all the joints on that timber. The arcs at the end of the posts wouldn't be found later, as the wood under them would be cut away for tenons. The old master builders planned it that way, so that others couldn't discover their tricks.

The timbers on this barn were small enough so that the next operation took place on a workbench inside the 1750 Dutch barn that functions as Richard's shop in Hancock. They snapped lines across all four sides of the timber at each location of a joint, plumb or at right angles to the long lines. It didn't matter which side the joint was on, squaring always started from the best side. Then they rolled the timber to the left and to the right and all the way over. The line on the last side simply connects the two from the left and right.

On outside wall posts other than the corner posts, the lines were centered. On corner posts and on other faces of posts, the lines were set off to accommodate the framing. That is, one side of the plumb line became one side of the framing—a mortise hole, for instance. Those plumb lines would remain through the framing. In the old days the line was rubbed with charcoal. Either charcoal or chalk will disappear in time.

For framing, they then had to draw a line on the second and third sides, parallel to the best face. Because the post was on a bench, they simply turned it best side down and drew a line parallel to the bench surface. Otherwise they would have had to mark the location with the compass and snap a line. This line shows the other side of the "ideal beam" that lies within the irregular one. It is that line, the plan line, on which the framing is based for that side.

They marked locations for the mortises with a pattern that size, laid along the new line at the point of the first squaring of the timber. Then they rolled the timber to the second best face, to mark the framing locations. Similar lines are

169

drawn to outline, for framing purposes, the outside of cross timbers, starting from the best face. They marked braces the same way.

Richard and Henry used the first plumb line on the best side to square off. When framing members meet at right angles, the line on one member needs to be extended to meet the plan line on the other member. Richard walked his compass three turns down the long line from its intersection with the plumb line and, starting at the same point, four turns down the line connecting the two pieces, marking each end point. Then he snapped a center line on a piece of scrap lumber and laid out five turns of the same setting. He marked beginning and end by driving a nail part way. This piece of scrap was now a tool, to be kept to prove all right angles on this job and perhaps others. Richard's grandfather had had one of these with his tools. He called it a "tryon," "triset," or "patronset."

As posts were always a bit out of square, Richard added a set amount to the length of the cross beams so that any gap could be squared. The recess or mortise framed into the post and the amount to be removed from the tenon he figured from the plumb line on the the post. This approach is like the "scribing" used by some builders who were not initiated into the art of squaring timbers with line and compass, laying out one beam on top of the other to mark the cuts. But for Richard, the word "scribe" doesn't mean tracing, it means "squaring" using the three-four-five right triangle. He thinks the term comes from the teachers or doctors, learned men, in the Bible, rather than from using a sharp instrument to etch a line.

Richard explains that the standard size of the present builder's square was based on earlier squares, which had no numbers on them, but could be used instead of a line to carry the plumb mark around four sides of a timber. The thickness of the square was the standard width of a mortise, as an aid in laying them out. Henry squared off the mortises with a chisel. The mortises were always cut before the tenons, which are easier to adjust, Richard notes.

But on this barn, Richard and Henry carefully formed the tenons, too, the old way, eschewing a saw for a chisel. "We did everything the old way, carefully, by hand, just as though we were framing that barn 250 years ago," Richard says.

The square rule, as it's applied today, is similar in principal, except for the numbers on the square and the fact that the wood comes sawn square from the lumber yard, so that the long lines don't have to be snapped. The edge of the wood is the line. But the same three-four-five triangle can be used for checking the square of large timbers. The goal is the same: when the timbers are erected in the frame, they should come together plumb, the posts rising at right angles to the sill, the plates at right angles to the posts, and so forth. A frame out of plumb is not as strong as a frame that is plumb. A builder will, therefore, wrench the frame to get it in shape. A way of checking the workmanship, then, is to see if the builder cheated, opening gaps in the joints while squaring the building.

Richard points out that many an old master builder put on the lines in such a way that they would disappear, because they would have given away the secrets of his trade. Some, however, left arcs and plumb lines, which Richard has found in his work. He notes that because hand-hewn beams are always a bit out of square, parts are not interchangeable. Using sawn timber, the builder could simply lay out four identical corner posts. Braces would fit anywhere. In the older ways, each joint was marked with corresponding Roman numbers. These the master builder left on, because they were used first in trying out a joint and later in the actual erection.

The braces on the small barn were customized but, using the scribe rule techniques, it wasn't necessary to lay them on the beams. Richard already had the plan lines on the posts and plates. Once he had measured the distance in on both, by the three-four-five triangle he knew how many turns long each brace should be. He cut the shoulder of each tenon at 45 degrees. The tenon itself was a right triangle, at right angles to the shoulder, left when he cut away another right triangle at each end.

Richard and Henry used the scribe rule to make the rafters. First they found the center point of the structure, "half your width," by walking the compass down the cross beams—when they were at full length, before the tenons were cut. The center point is that from which an equal number of turns reaches each end. They then made a right angle off the center line by the compass method. Using the same compass setting that determined half the width of the structure, they walked one turn from the center and marked that point. Then they walked one turn and marked in the opposite direction. They spread the compass to touch the two points. This setting, walked along the timber that will become the rafter the same number of turns as it took at the prior setting to find the half width, equals the length of the rafters. They cut them at right angles.

The way to visualize the geometry is to draw a half circle. The diameter is the cross beam. A right angle forms a radius. Rafters are chords connecting the radius and diameter at the circumference. Who figured out that rafters could be laid out that way and how long ago? Was it the Greeks or the Egyptians? "This is an important part of the scribe rule," Richard says, so perhaps the skill derives from Biblical times.

Each pair of rafters was joined with a small mortise and tenon cut plumb to the rafter, possible because the rafters joined at right angles. The little barn had no ridge pole. The bottoms or tails of the paired rafters were laid out upon the cross beams at the proper width. Then a line was snapped in such a way that it marked the angle at which the rafter would cross the plate, and the extent to which the plate would require cutting a notch out of the rafter. In the scribe rule method, used on this barn, the rafter is locked behind the inside of the plate. In another method, the tail might extend beyond the outside of the plate to form overhanging eaves. When installed, wooden pegs help hold the tail of the rafter to the plate.

Framing, the art of laying out and cutting the joints, in most cases takes the same amount of work and time, regardless of the size of the barn. An exception would be Dutch barns, in which enormous tenons protrude through mortises and are wedged. The small barn followed the standard rule, though: it took the two men about a day to do each post, or three to four days for each bent. Framing the entire small barn took twelve days.

These figures provide a measure of the difference between timber framing and pole construction, in which two men using power saws could put up the studs, sills, and plates in one day. Timber framing is labor-intensive. It is also strong, beautiful, and a statement of history.

Richard drove his flatbed truck, loaded with beams, to the site, at the junc-

tion of Routes 2 and 7 in front of the Williams Inn, on Friday afternoon, September **Volunteers raise a wall section of the small barn by hand. (Cheryl Boyer photo)**

25—exactly forty years from the first opening of the '53 House. The truck got stuck in the poorly drained ground at Field Park, but that didn't make any difference until it was time to go home.

He knew it was going to be a tall order to raise even a small barn in one day, using for help anybody who showed up, so he arrived an hour early— 7 a.m.—Saturday, September 26, a lovely, warm autumn day. The next person to show up, David Boyer, a rugged, tall, off-duty Williams College Security Officer, gladened Richard's heart. He knew now someone would be able to provide the oomph when needed. Size is helpful, too, so Richard was glad when a tall Williams student, Lee Osbaldston, happened by. Then Eric Youngquist and Charles Schlesinger appeared, as well as Henry N. Flynt and Lauren R. Stevens.

Cheryl Boyer sits on the shoulders of her husband, Dave, to drive in a wooden peg. The erection of the barn celebrated the 40th birthday of the 1753 House, in the background, built to commemorate Williamstown's 200th birthday. (Lauren R. Stevens photo)

They spread some planks on which to rest the sills. Then they began to put together a side, including the sill. The braces were joined where they crossed each other, as well as joined to the posts and plate.

By this time, enough people had arrived to raise the east side and hold it in place while Richard fashioned temporary braces from slender birch trees. Raising an entire side with sill was convenient for a project without any foundation, but not the same as the more normal approach of raising by bents—two posts and the crossbeam. Richard told the workers that "the most important section of timber to raise is always the first one."

Richard hustled to fit together the west side. While the pieces had gone together at Hancock shop, he was gratified to discover they still did before a crowd. Nevertheless certain joints needed a few smart raps from Gramp's "Gentle Persuader"—or its descendant. The six-man-team raised the west side, holding it in place while Richard fastened more young trees to make more temporary braces.

David Boyer's wife, Cheryl, climbed onto her husband's shoulders to drive in the trunnels or pegs, while their daughter, Chelsea, watched.

Meanwhile, other folks arrived to set up a refreshment stand in front of the '53 House. Town Selector Anne R. Skinner cut a birthday cake. Punch and cake were courtesy of The Williams Inn, which faces onto Field Park. Guests came by, while the equally curious gawked from their

Richard and a volunteer haul a rafter to the plate of the small barn that was framed the old way. (Cheryl Boyer photo) ➤

174

Eric Youngquist helps Richard install a few shakes to show how the roof would have been put on. (Cheryl Boyer photo)

automobiles, headed for Vermont on U.S. Route 7.

Under Richard's direction, the tenons of a cross beam were fitted into beams on both sides. A piece of rope helped hold some beams in place while they were fitted. Richard laid planks across the sill for workers to stand on while they maneuvered the cross beams in place.

By this time, the middle of the day, people were coming as observers, learning about the '53 House, barn raising, and the punch. Richard talked a group of passing college students into helping, but they disappeared soon after he turned his attention elsewhere.

Crew members clambered onto the top of the frame to start driving the pegs that would hold the joints securely. The handmade ladder used to reach the loft in the '53 House turned out to be the right length to reach the plates in the barn, which made it easier to carry up some planks to spread across the plates so that workers would have a place to stand while erecting the rafters.

The afternoon was quickly passing. Some of the workers had to go, yet Richard was determined to get the rafters in place. The first pair was brought up individually, placed in the middle position, pegged to each plate, and braced with a board nailed to to the west plate. Then a peg was driven through the mortise and tenon at the ridge.

Then other pairs of rafters were installed, moving out from the center one in both directions. Pieces of scrap lumber kept the rafters upright. The sun lowered; hunger and fatigue began to gnaw

Chelsea Boyer was a faithful spectator. (Cheryl Boyer photo)

at the workers. Yet a Daylight Savings Time sun bathed the project in light.

With five pairs of rafters in place the volunteers thought their work was over, yet Richard had it in mind to nail a few shakes in place, just to show how it was done. He fastened three poles between two sets of rafters, tacking three shakes on the bottom row, two on the second row, and one at top, the row above covering the gap between two below.

The barn stayed until the next weekend, when a smaller crew disassembled it. Those six shingles were the only roof that small, carefully wrought demonstration barn ever had, as its timbers were also inside the Great Saratoga Barn when it burned.

EPILOGUE

IT ALL COMES DOWN TO learning about trees, Richard thinks. When he was growing up, Gramp took him into the forest and taught him to tell one tree from another. Gramp made a game out of it. Richard had to be able to identify them with leaves and without; when the trees were standing and when they were chunked up for firewood. It was important to know trees and what he could do best with each species.

The fancy word for this knowledge is *dendrology*—the scientific study of trees. It consumes a major part of a student's time in forestry school. That study is valuable for anyone who wants to make her living from the woods, but that is not exactly what Gramp had in mind. To Richard's grandfather and to Richard, knowing trees is just a basic living skill, not specialized knowledge. It ought to be second nature.

For Gramp, knowing trees was part of being a farmer and a carpenter. For Richard it is the basic skill of barn restoration, like knowing how to skate so well we don't have to think about it when playing hockey. Of course, in other lines of carpentry it is possible to go to the lumberyard to order 12-inch, clear white pine or 2 x 6 spruce. Richard needs to be able to know that the beam with one end rotted is white oak and, if necessary, to be able go into the woods to cut another one to match.

"I think children need to be introduced to the forest firsthand, just as they should be introduced to the past," Richard says. Virtual reality is not the same as reality; computer simulation thankfully is not a substitute for cellulose. Books are great and instructive, but even books are not the whole thing.

"My advice," says Richard, "is take your children—one at a time is best—to the woods, or a park if you have to. Get a book, if necessary, and just learn the trees. This is a start toward saving the past."

What else does he recommend? "You're probably not going to be able to examine eighteenth-century homes or seventeenth-century barns," he says, "but try." Richard was blessed with a location where he could see a combination of old barn types and old houses, the way they were, in use. The authentic, the real. Looking at old barns now turned into the heart of a house or a living museum, "we can see a truth and we can imagine those old timers working at their trades."

"I can take down a tree in the old way and frame it in the old way," Richard says. "That needs to be re-lived."

Yes. But now Richard has a heart he must pay attention to. His ankles are still weak from the lead poisoning. He lost three precious barns in one fire, the timbers of two inside the standing barn. As his son, David, notes, Richard is at the point in his career, recognized nationally for his work, that he ought to be able to pick his jobs and command top dollar, yet he still has to hustle and certainly is not paid commensurate to his skill—or his reputation.

"Not many people are bold enough to take down an ancient barn," David says. He mentions a couple of other barn restoration firms that operate on a different philosophy. They base their work on the old barn but re-erect it with new parts, today's technology, even laminated beams. A case can be made for that approach. The barn itself then becomes kind of history of barns through time, instead of a historical artifact arrested at a certain point in time. But using sawn wood or glued wood to replace hand-hewn beams is not Richard's way.

Of those courageous enough to tackle barn work to begin with, only a handful will tell a client that the barn must be replaced as close to its original form as possible. That is Richard's way. Generally, David points out, even if the client is interested in authenticity, he still doesn't want to pay the full freight. "It's just an old barn," even an intrigued client thinks. David's view is that people ought to be paying his father and himself as though they were creating a work of art, not just moving "an old barn."

Richard is not sure what to do. Filling in the empty spaces in his museum would be a huge job, even if he were younger, and it would require a lot of money. He wonders if he could run a school for apprentice barn builders, which is after all what he has done all his life, but he's not sure about the number of people who would pay to learn those skills.

Most of all, he wants to save ancient barns, and he wants people to come to understand their value. Richard Babcock believes in a divine plan. He was placed in Hancock, Massachusetts, in the second half of the twentieth century, to save the heritage of barns. The sequence of his life was arranged to that end. But sometimes, it is hard for him to know exactly what the next move will be.

■ ■ ■

Recently Richard walked through the blackened ruins of his burned barns, where he doesn't often go. He was considering what he had left and whether anything could be raised from the ashes. He poked with his foot at a large beam from the French barn, knocking off some charcoal.

Suddenly curious as to how far into the oak the fire had reached, he cranked up the chain saw. He removed a slice from one end and examined it.

The charring, he immediately saw, wasn't more than one-half inch to an inch deep. He could count nearly all of the growth rings. The white oak was about 110 years old when it was cut. The rings indicated that during its first decades the tree had grown quickly and then it slowed down.

The original timbers were larger than they needed to be. It occurred to Richard for the first time that he could shave off the charring, resize the beams, add in pieces to replace those destroyed, and raise the essential Norumbega barn after all.

It is a glorious thought.

Let's continue to go to the barn.

GLOSSARY

Terms Related to Ancient Barns

English Barn.

Rectangular shape, with large doors on the sides. All have handwrought metal hinges. Small girts between the posts provide a nailing place for the vertical siding. Built of oak. One of the first barn types built on the coast of what was to become the United States and later elsewhere in New England.

All have distinctive gunstock posts, made by turning tree upside down to take advantage of the natural widening at the trunk and, in hewing, leaving the extra width at the top to house the tie beam and the plate. The tie beam runs over the wall plate with a half dovetail bottom connection. At the bottom of the post, a single tenon fits into a mortise in the sill. Some have major rafters which are housed in the tie beams, forming a truss with a collar tie or two purlin legs between the rafters and the tie beam. Another roof system has purlins framed into these trusses at a center point in the major rafters. The purlins go from major rafter to major rafter on both sides of the roof. The secondary rafters are set over the purlin beams. A form of roof had a ridge beam, either with or without purlins.

Dutch Barn.

Generally square and large, these were first built along the Hudson and Connecticut rivers at the same time as the English barns on the coast. The Dutch barns are larger than the English and built of pine, except for the sills and plates, which are generally oak. The large doors grace the gable ends, for the eaves of the roof are closer to the ground than on the English barn. The doors, small and large, hang on wooden hinges. A small roof protects the doorways from the weather. Wall posts every four feet serve as nailers for horizontal siding.

The distinctive H bent includes the huge anchor beam, two posts and heavy braces. Protruding tenons on the anchor beams, rounded on the ends, pass through the posts and are fixed with tapered oak wedges. The heavy plank floor have splines between the planks. The planks meet on special ledges on center and purlin sills. The area open to a wagon is twice the size as in an English barn. Dove holes high in the gable ends help take off the moisture in the winter and the heat in the summer, and dry the underside of the roof. Purlins for the roof system attach to the anchor posts. The rafters meet with mortise and tenon at the ridge and are pegged into the outside wall plates. The outside walls are well beyond the anchor posts.

German Barn.

Built of oak. Rectangular, with less roof than the Dutch barn has. Girts between the posts provide nailers for the vertical siding. The doors are on the sides.

Characteristically German barns have a swing beam, a long span that keeps the floor clear of center posts so the farmer can swing a team around under it. This truss is formed by a smaller beam on top, braced and connected to the larger piece to hold it up at the center. Some German barns have several swing beams, but one is always the largest. That swing beam has square mortises on one face to receive the poles that hold the wheat. Another type of German barn has anchor beams, like the Dutch, built of oak. The protruding tenons are square. The outside wall receives vertical siding but the siding runs horizontally on the gable ends, where folding doors also swing on wooden hinges.

Scottish Barn.

Similar in shape and size to the English barns, with doors on the sides. Scottish barns use posts that taper their entire length, not the distinctive gunstock shape. The tie beam is supported by the wide top of the post but connected to the wall plate by a tenon laid flat and pegged. It has to be held above the post and dropped into place—a kind of secret joinery. In some of the Scottish barns the centered tie beam connects to a lower point on the posts, while on the end wall tie beams connect directly to the plate. The outside wall plates jut out two inches over the posts. A groove on their underside receives the siding. Narrower Scottish barns have no purlins, while the wider do. Some have ridge beams.

Anchor Beam. Large (1'x2' cross-section) cross beam in Dutch barns, with large braces.

Auger. A tool with an iron shank and a wooden handle used to drill holes.

Beams. Logs hewn to flat sides, fastened together with mortise and tenon and pegged, to frame a barn.

Bent. The basic unit for timber-frame assembly—vertical posts, horizontal beams, and braces.

Block and Tackle. Two sets of pulleys with the line running between them, used with a gin pole.

Brace. A smaller piece of wood used to stiffen corners to check the sway where two beams meet.

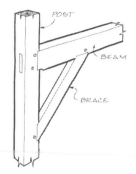

Bull Wheel. A vertical shaft housed in a timbered framework, turned by pushing on poles stuck in slots. Increases leverage. Used to take up the line on block and tackle or for moving barns. Similar to a capstan on a ship.

Carpenter's Compass. The basic tool of old timber framing. The distance between the legs of these dividers can be adjusted.

Collar Tie. Short beam that fastens paired rafters together.

Cross Beams. A beam running horizontally between two posts.

Cruck. Naturally curving timber; pairs act as posts and principal rafters in cruck framing.

Dry Stone. A stone wall laid up without mortar.

Dove Hole. Openings in the upper gable that allow birds access to barns and, more importantly, provide ventilation.

Dovetail. A type of joint in which the tenon is cut in and then widens as it leaves the beam. The mortise is shaped to receive it, with a wedge to lock it.

Frame. Used as noun, the beams that support a barn. As a verb, it means to measure, cut, and fit together the beams that make up the frame.

Folding Door. A door that opens on hinges.

Froe. A blade with a detachable handle used to split shake shingles or clapboards.

Gable. The section of wall under the slanted portion of the roof.

Gin Pole. A slender tree, fitted into a base and supported by guy lines, used to lift and lower beams.

Girt. A short, horizontal beam, used in an intermediary position to connect two posts or beams.

Gunstock. The characteristic enlargement at the top of the posts in an English barn, used to receive cross beams and plates.

H. The heavy posts, anchor beams and braces used to form the heart of a Dutch barn. The purlins pass over the tops of the posts.

Hay Fork. A mechanized device that runs along a track at the ridge of a barn. To install it, farmers removed collar ties and other framing members.

Hewn. Beams shaped by an ax or adze; as opposed to sawn.

Joinery. The methods of fastening two pieces of wood together based on shaping the wood and pinning rather than nailing.

King Post. A post mounted on a cross beam in the center, fastened to it in such a way that it holds the cross beam up. When it, in turn, is supported by long braces that run from its top, at the ridge, to the plates, it forms a **king post truss.**

Line. String or twine, rubbed with charcoal or chalk, pulled tight and snapped to leave a mark; also a rope used in a block and tackle.

Loft. The space immediately below the roof.

Mortise. In joinery, the hole that receives the tenon.

Mow. The part of the barn where the hay or grain is stored.

Peg. A round oak pin, 1–2" in diameter, 6–12" in length, used to lock joints.

Pentice Roof. In Dutch barns, a small roof to keep rain from rotting a doorway. The pentice is a special rafter that is pinned to the barn at one end. Three of them hold up the pentice roof.

Plate. Horizontal beam at the top of the posts along the edge of a building; receives the rafters and ties the sections together.

Plumb. Vertical to the horizon. A plumb is used to prove verticals when line and post are parallel. A plumb bob is a weight hanging on a line. When it is directly under the apex of a triangle, the base of the triangle is level.

Pole. Slender tree mortised into beams and used to lay wheat on to dry. *Or,* Sawn dimensional lumber used in modern construction. A pole barn would be framed with 2x4s, 2x6s, 2x8s, etc.

Post. A vertical framing beam.

Post and Beam. A frame made of vertical and horizontal members fastened with joints.

Purlin Plate. A beam running horizontally to support rafters at their midpoint and tie the sections together.

Queen Post. Part of a truss, used to support a roof and cross beam.

Rafter. Used to carry a roof, it runs from plates to ridge.

Roots Barns. Barns built by early settlers in their manner of their native country.

Ridge Beam. A beam that receives the tops of the rafters; since most old barns did not have a ridge beam, the line formed by the upper end of the rafters is called the ridge.

Saltire. In French barns, braces tenoned into plate and posts, sometimes half-lapped into girts.

Score. In hewing, cutting into the side of a log at an angle to loosen chips.

Scribe Rule. A set of ancient rules used with a line and carpenter's compass to frame timbers.

Sill. Horizontal beam on top of the foundations that receives the posts and flooring system. A center sill runs down the middle of the floor. Sometimes rotted sills have been replaced with rocks but all barns had sills when originally erected.

Silo. Cylindrical-shaped building to hold corn or grass with minimum air for feeding cattle during the winter.

Slick. Large, flat blade, never struck with a hammer but pushed by hand as a plane to smooth a tenon.

Square Rule. The modern term for framing according to rules, using lumber squared at mill.

Stud. Upright member of the wall framing.

Swing Beam. In German barns, a beam large enough to make center posts unnecessary, leaving an area to swing a team of horses into their stalls.

Tenon. Projection at the end of a timber that fits into a mortise of another timber. Usually the joint is pegged together.

Threshold. Board placed across the bottom of door frame to prevent threshed grain from escaping and the rain from splashing in when the door is closed.

Tie Beam. Beam that joins two vertical members or rafters.

Timber Frame. To build with posts and beams.

Tithe Barn. Large barns built during the Middle Ages to store 10 percent of a harvest, which landowners were obliged to donate to the church.

Trunnel. From "treenail," a wooden peg, usally oak, with a tapered head and wedge inserted in the opposite end.

Truss. Combined vertical and horizontal members to strengthen a horizontal beam.

BOOKS ON BARNS AND RELATED HISTORY

Alphonce, Jean. *La Cosmographie*. A copy of the 1545 manuscript is in the Harvard University Library (Paris: E. Leroux, 1904), from the original in La Biblioteque Nationale.

Arthur, Eric and Dudley Witney. *The Barn: A Vanishing Landmark in North America*. Greenwich, Conn.: New York Graphic Society, Ltd., 1972. Beautiful images emphasizing Canadian barns.

Babcock, Richard W. *The Barns at Wolf Trap: A History of the Barns and Their People*. [Hancock, Mass.], 1982.

_____. *Barns of Roots America*. [Hancock, Mass.], 1989.

_____. *Barns in the Blood: Master Builder, Discover*. [Hancock, Mass.], 1993.

Burden, Ernest. *Living Barns: How to Find and Restore a Barn of Your Own*. New York: Bonanza Books, 1977. Discusses Richard's contributions including the art of moving a barn; photographs of conversions.

Endersby, Elric, Alexander Greenwood, and David Larkin. *Barn: The Art of a Working Building*. Boston: Houghton, Mifflin, 1992. Gorgeous photographs by Paul Rocheleau and helpful text by the New Jersey Barn Company founders.

Fitchen, John. *The New World Dutch Barn*. Syracuse: Syracuse University Press, 1968. A model of research into the literature.

Goodell, William. *The American Slave Code*. New York: American and Foreign Anti-Slavery Society, 1853.

Hakluyt, Richard. *The Principal Navigations, Voyages, Traffiques and Discoveries of the English Nations*. Edited by Edmund Goldschmidt. Edinburgh: 1889. Includes passages from *Divers Voyages Touching the Discovery of America* (1582) and *Principal Navigations* (1589)—the only version that includes Ingram's trip.

Halsted, Byron D., ed. *Barns, Sheds and Outbuildings: Placement, Design and Construction.* Brattleboro: Stephen Greene Press, 1977, Reprint of 1881 original. Urges silos and owner-built pole barns on the farmers of his day.

Howe, Henry E. *Prologue to New England.* New York: Farrer and Rinehart, 1943. A classic history of early European exploration of this continent.

Kettle, Walter A. *The Palatine Emigration of 1708–1709.* Philadelphia: Dorrance & Co., 1937. Reprinted Baltimore: Genealogical Publishing Co., 1965.

Klamkin, Charles. *Barns: Their History, Preservation and Restoration.* New York: Hawthorne Books, 1973. Finds old barns have "a feeling of essential rightness."

Leach, Douglas Edward. *Colonial Frontier, 1607–1763.* New York: Holt, Rinehart and Winston, 1966. History as the frontier moves through upstate New York.

Matson, Peter H. *A Place in the Country: A Narrative on the Imperfect Art of Homesteading and the Value of Ignorance.* New York: Random House, 1977. A somewhat patronizing attitude toward Richard, the "barn mover" referred to anonymously.

Niles, Grace Greylock. *The Hoosac Valley: Its Legends and Its History.* New York and London: G.P. Putnam's Sons, 1912. An intriguing example of the genre of local history, with its strengths and weaknesses.

Noble, Allen G. and Richard K. Clee. *The Old Barn Book: A Field Guide to North American Barns and Other Farm Structures.* New Brunswick: Rutgers University Press, 1995. The most comprehensive study, organized like a botanical guide. Appears to benefit from Richard's studies indirectly, especially concerning German barns.

O'Callaghan, Edward B. *Documentary History of New York.* 4 vols. New York: Weed, Parsons & Co., 1849.

Perry, Arthur Latham. *Origins in Williamstown.* New York: Charles Scribner's Sons, 1894. Local history by a professional historian.

The Postal Service Guide to U.S. Stamps, 9th edition. Washington: United States Postal Service, 1983.

Reed, Max. *The History of the Mohawk Valley.* New York: G.P. Putnam's Sons, 1901.

Russell, Howard S. *A Long, Deep Furrow: Three Centuries of Farming in New England.* Hanover, NH: University Press of New England, 1976. Provides invaluable information on the first three centuries, up to World War II.

_____. *Indian New England Before the Mayflower*. Hanover, NH: University Press of New England, 1980.

Schuyler, George W. *Colonial New York*. 2 vols. New York: Charles Scribner's, 1885.

Shaw, Edward. *Civil Architecture*. Boston: [1832]. Writes about the square rule and scribe rule.

Sloane, Eric. *American Barns and Covered Bridges*. New York: Funk and Wagnalls, 1954.

_____. *An Age of Barns*. New York: Funk and Wagnalls, n.d. The better of the two Sloane barn books, complete with his enthusiasm and fine sketches.

Sobon, Jack A. *The Scribe Rule or The Square Rule: Traditional Timber Frame Layout Systems*. [Windsor, Mass.], 1994. One of Richard's protegés attempts to sort out a technical issue.

Sylvester, Nathaniel Bartlett. *History of Rensselaer Co., New York*. Philadelphia: Everts & Peck, 1880.

Thevet, Andre. *La Cosmographie Universelle*. Paris: P. L'Huilier, 1575.

Valentine, T.W. *Valentines in America 1644–1874*. New York: Clark and Maynard, 1874.

van Loon, Hendrich. *Life and Times of Pieter Stuyvesant*. New York: Henry Holt, 1928.

Weise, Arthur James. *The Discoveries of America to 1525*. New York: G.P. Putnam's Sons, 1884.

Wheeler, Jacob D. *Wheeler's Law of Slavery*. Dating to 1837, this book was used by slaveholders to determine their "legal" rights.

Whittier, John Greenleaf. *The Works of...*, Vol. I. Boston and New York: Houghton, Mifflin and Co., 1848.

Wilford, John Noble. *The Mapmakers*. New York: Alfred A. Knopf, 1981. Tells of Mercator's 1569 map, which shows Norumbega.

Ziegler, Philip. *Storehouses of Time: Historic Barns of the Northeast*. Camden, Maine: Down East Books, 1985. Author calls Richard his mentor.

INDEX

About the Authors

Harding-Glidden

Judith Monachina

Richard W. Babcock lives among his barns in Hancock, Massachusetts, but travels extensively to inspect, dismantle, and rebuild old structures. **Lauren R. Stevens** has published five books on environmental topics.